Audrey D. Levine

35⁰⁰
counter box

Best Practice Guide

on the Control of Lead in Drinking Water

Metals and Related Substances in Drinking Water Series

Best Practice Guide
on the Control of Lead in Drinking Water

Edited by

Dr Colin Hayes

Swansea University, UK

IWA Publishing

London • New York

European Science Foundation

SETTING SCIENCE AGENDAS FOR EUROPE

cost

Published by **IWA Publishing**
Alliance House
12 Caxton Street
London SW1H 0QS, UK
Telephone: +44 (0)20 7654 5500
Fax: +44 (0)20 7654 5555
Email: publications@iwap.co.uk
Web: www.iwapublishing.com

First published 2010

Typeset in India by OKS Prepress Services.

Printed by Lightning Source.

British Library Cataloguing in Publication Data
A CIP catalogue record for this book is available from the British Library

Library of Congress Cataloging-in-Publication Data
A catalog record for this book is available from the Library of Congress

ISBN 10: 1843393697

ISBN 13: 9781843393696

Contents

Authors

Dr Colin Hayes, Swansea University (UK)
Roger Aertgeerts, World Health Organization (Int)
Lisa Barrott, MWH (UK)
Angelika Becker, IWW (D)
Maria Joao Benoliel, EPAL (PT)
Dr Brian Croll, Consultant (UK)
Professor Marc Edwards, Virginia Tech (US)
Dr Daniel Gari, National Institute of Public Health (CZ)
Dr Eddo Hoekstra, Joint Research Centre (EC)
Martin Jung, Austrian Institute of Technology (AT)

Dr Adam Postawa, AGH Krakow (PL)
Achim Ruebel, IWW (D)
Dr Larry Russell, Reed International Ltd (US)
Michael Schock, US Environmental Protection Agency (US)
Nina Skubala, Swansea University (UK)
Professor Stanislaw Witszak, AGH Krakow (PL)
Marcel Tielemans, Het Waterlaboratorium (NL)
Dr Magdalena Zabochnicka-Swiatek, Technical University of Czestochowa (PL)

Acknowledgements

This Best Practice Guide derives from the knowledge assembled by the European research network COST Action 637, supported by a wide range of experts from 26 European countries, the US and Canada. The funding received from COST over the period December 2006 to November 2010 is duly acknowledged.

The main objective of COST Action 637 is to stimulate better control of metals in drinking water and to minimise environmental and health impacts. The Action aims to contribute to the implementation of the European Drinking Water Directive and to the Environment and Health programme of the EU Commission and the UN/WHO Protocol on Water and Health. The Joint Research Centre of the European Commission provides scientific input in the COST Action but is also the link to the European Commission. COST stands for European Cooperation in Science and Technology and is the oldest and widest European inter-governmental network for cooperation in research. COST is supported by the EU RTD Framework programme (www.cost.esf.org).

Thanks are also due to all the water companies, regulators, health agencies and research institutes who have provided access to data.

Acronyms

AL	Action level
DAC	Daily average concentration
DIC	Dissolved inorganic carbon
E	Equilibrium concentration
EU	European Union
LCR	Lead Copper Rule (US)
M	Initial mass transfer rate of lead dissolution to water
RDT	Random daytime sampling
UK	United Kingdom
US	United States
WHO	World Health Organization
30 MS	30 minutes stagnation sampling

Definitions

Compliance monitoring	Monitoring at the point of compliance to verify that water supplied for human consumption is in compliance with its quality requirements.
Inventory monitoring	Monitoring to estimate the lead problem in a water supply zone or country.
Operational monitoring	Monitoring activity to check the quality of source water, and to validate the operation of the water treatment plant, the distribution network and the domestic distribution system up to the tap.
Plumbosolvency	The tendency for drinking water to dissolve lead.
Water safety plan	A Water Safety Plan is a comprehensive risk assessment and risk management approach that encompasses all steps in water supply from catchment to consumer. It comprises as a minimum the three essential actions that are the responsibility of the drinking-water supplier in order to ensure that drinking-water is safe. These are: a system assessment, effective operational monitoring, and management.
Zone (zonal)	An area of water quality assessment, either an entire or a part of a water supply system.

About this Best Practice Guide

This Guide is one of a series produced by the International Water Association's Specialist Group on Metals and Related Substances in Drinking Water. It is a state-of-the-art compilation of the range of scientific, engineering, regulatory and operational issues concerned with the control of lead in drinking water. It will be of interest to water utility practitioners, health agencies and policy makers.

The Specialist Group is supported by members from 26 European countries, Canada and the United States. It is an active research network and has regularly convened international conferences and seminars. It has close working links with the World Health Organization, the European Commission's Joint Research Centre, Health Canada and the US Environmental Protection Agency. The Specialist Group developed out of COST Action 637 (www.cost.esf.org), a European research network.

The Guide is supported by a two-day training course and an abbreviated version of the Guide aimed at small community water suppliers. Information about training, the guide for small community water suppliers and the research network in general is available from www.meteau.org

Disclaimer

Whilst every reasonable attempt has been made to present the information in this Guide in a fair and balanced manner, the reader should none-the-less satisfy themselves of its relevance to their specific circumstances. It must also be appreciated that some aspects of the topic of plumbosolvency control do not enjoy total consensus of opinion and that practices have varied around the world. The designations employed and the presentation of the material in this publication do not imply the expression of any opinion whatsoever on the part of any organization mentioned in this publication concerning the legal status of any country, territory, city or area or of its authorities, or concerning the delimitation of its frontiers or boundaries.

The mention of specific companies or of certain manufacturers' products does not imply that they are endorsed or recommended by the publisher or authors in preference to others of a similar nature that are not mentioned. All reasonable precautions have been taken by the authors to verify the information contained in this publication. However, the published material is being distributed without warranty of any kind, either express or implied. The responsibility for the interpretation and use of the material lies with the reader. In no event shall the publisher or authors be liable for damages arising from its use. The views expressed by the authors do not necessarily represent the decisions or the stated policies of any organization referred to in this publication.

© 2010 The Authors. *Best Practice Guide on the Control of Lead in Drinking Water*. Edited by Dr Colin Hayes. ISBN: 9781843393696.
Published by IWA Publishing, London, UK.

Foreword

by Roger Aertgeerts, World Health Organization

Lead has been a challenge and a bane for water suppliers since historical times. The statement that the extended use of lead pipes in aqueducts was one of the contributing factors to the fall of the Roman Empire has entered collective memory. The numerous articles printed in leading scientific journals, in the United Kingdom and United States, in the late nineteenth century, documenting thousands of cases of lead poisoning caused by lead water pipes, have largely faded in the mist of history. These cases often resulted in death, paralysis, blindness, insanity, convulsions, miscarriages and stillbirths.

One historical example in particular gives pause for thought. In the summer of 1887, the town of Bacup, in England, experienced a severe drought. The town's water reservoir went dry for several weeks and without water regularly running through the distribution systems the interior of the lead service pipes were exposed to air. This caused the encrustation of organic and inorganic compounds that had built up over the years to break away from the pipes. Once the protective barrier was gone and water was returned to the pipes, water began to dissolve the lead. Shortly after, 404 cases of water-related lead poisoning were documented by Bacup's health office. With climate change induced droughts making uninterrupted water supply an increasingly difficult challenge in many countries, the story of Bacup looms as an early warning whose lesson should be learned earlier rather than later.

The World Health Organization has progressively tightened its guideline value for lead from a maximum allowable concentration of 0.1 mg/l to the current 0.01 mg/l. The WHO Guidelines recognize that lead is exceptional, and that most lead in drinking-water arises from plumbing, and the remedy consists principally of removing plumbing and fittings containing lead. As this requires much time and money, it is recognized that not all water will meet the guidelines immediately. Meanwhile, all other practical measures to reduce total exposure to lead, including corrosion control should be implemented. Treatment to reduce plumbosolvency usually involves pH adjustment and, additionally, dosing with ortho-phosphate may be necessary.

Independent studies in the European Union have estimated that about 25% of houses may have lead piping, potentially affecting 120 million people. Based on the age of the housing stock, the situation in the eastern part of the WHO European region can be expected to be comparable or worse. Problems with lead in drinking water have also necessitated corrective action in many parts of the United States and the position in Canada is under active review. It seems that lead piping and the ensuing problems for drinking water are a common heritage in many of the industrialised and urbanised nations.

Twenty-four countries of the WHO European region have ratified the Protocol on Water and Health to the 1992 Convention on the Protection and Use of Transboundary Watercourses and International Lakes. Amongst the basic aims of the Protocol are the provision of safe drinking-water, and the reduction of water-related diseases. Under the terms of the Protocol, Parties are obliged to set targets and report on progress towards the attainment of the targets, usually every three years. The Task Force on Indicators and Reporting has identified lead as one of the core chemical parameters on which targets should be set and progress towards compliance with the WHO Guideline Value should be reported.

It is hoped that this Best Practice Guide will assist the countries of the WHO's European region and the Parties to the Protocol in particular, in reaching universal compliance with the WHO Guideline Value for lead. The Guide should also provide much helpful information on how to tackle the problem of lead in drinking water elsewhere in the World, even where different regulatory standards are used, such as in the United States.

Executive Summary

The main source of lead in drinking water is from lead service pipes and the lead pipes inside dwellings and in the older districts of some Cities and Towns, up to 90% houses may have a lead pipe. Problems can also be caused by lead leaching from devices, pipes and plumbing components made of brass, and from lead-containing solder.

Lead dissolution in low carbonate content waters can be significantly reduced by sufficient pH elevation. However, depending on carbonate concentration and other water chemistry factors, it may be insufficient to meet modern-day standards for lead in drinking water. Other characteristics of pipe scales are also important to plumbosolvency control optimization. Iron "red-water" discolouration will exacerbate lead dissolution from lead pipes and will need to be minimised if fully effective plumbosolvency control is to be achieved by corrective water treatment. Manganese and aluminium can also interfere. Lead can accumulate in premise plumbing pipe scales downstream of lead piping, particularly (but not only) old galvanized steel pipes, and even after the lead pipe is removed, continue to be released into the water to expose consumers to lead levels above contemporary standards. Organic matter, particularly humic and fulvic acids, will exacerbate lead dissolution from lead pipes and will need to be minimised if fully effective plumbosolvency control is to be achieved by corrective water treatment. Poly-phosphate dosing can also exacerbate lead dissolution from lead pipes.

Random daytime (RDT) sampling offers an unbiased approach for zonal compliance assessment but suffers from poor reproducibility, particularly with small sample numbers. Bulking RDT data over several years enables treatment changes to be adequately demonstrated, albeit allowance must be given to the response time of lead pipes and statistical tools can be used to determine levels of confidence in RDT sampling data.

Random stagnation sampling is not suitable for zonal compliance assessment because of dilution effects and stagnation sampling at selected properties, for the purpose of zonal compliance assessment, is sensitive to property selection. However, stagnation sampling can benchmark treatment changes at suitably selected houses and provide a direct measure of the reductions in plumbosolvency achieved.

The regulatory approaches in Europe and North America have been beset by problems due to difficulties associated with sampling for lead in drinking water. In Europe, there are proposals to normalise compliance monitoring by the use of random daytime sampling and a trend is emerging that will include risk assessment and risk management in the regulation of lead in drinking water. In consequence, water suppliers will in future need to look more deeply into the causes and extent of plumbosolvency problems in their area.

Lead is toxic to humans and lead poisoning is exhibited by a wide range of clinical conditions. Adverse health effects include: interference with haemoglobin biosynthesis; interference with calcium and vitamin D metabolism; gastrointestinal irritation; dullness; restlessness; irritability; poor attention span; headaches; muscle tremor; abdominal cramps; kidney damage; hallucination; loss of memory; encephalopathy; hearing impairment, gonad dysfunction, and violent behaviour. Lead can accumulate in bone and fatty tissue, with subsequent release, particularly during the latter stages of pregnancy. Most attention has been directed towards the retardation of child development, especially reductions in IQ. Lead in drinking water correlates with lead in blood in numerous studies and numerous studies correlate clinical effects with different concentrations of blood lead of which reductions in the IQ of infants are of greatest concern.

Virtually all drinking water is sufficiently plumbosolvent to cause exceedence of modern day standards for lead in drinking water, unless corrosion inhibitors are dosed and/or optimised pH adjustment is implemented. As an early estimate, about 25% of houses in Europe are possibly supplied by a lead pipe, and between 2 and 3% in North America. On the basis of two major European studies, two-thirds of the houses supplied by a lead pipe will likely fail the WHO Guideline Value of 10 μg/l for lead in drinking water.

At the zonal level (e.g.: a City or Town), risk from lead in drinking water can be defined by compliance with the WHO Guideline Value of 10 μg/l, on the basis of random daytime sampling. Computational modelling predicts that zones with a moderate plumbosolvency and more than 30% of houses with a lead pipe will require zone-wide action and zonal failure profiles give a better insight into likely health significance. The individual houses at most risk will have a lead pipe at least 10 metres long if the internal pipe diameter is 12 mm.

The simplest way to estimate the occurrence of lead pipes is housing age, from which plumbosolvency maps can then be drawn showing those areas at greatest risk, but more detailed inspection surveys are recommended. There are six basic technologies in use today for lead service pipe rehabilitation, namely: (i) open-trench, replacement along existing route, (ii) replacement along a new route, (iii) pipe pulling of a new pipe, (iv) slip lining, (v) pipe coating, and (vi) internal cathodic protection. Costs normally vary from €250 to €2,000 per lead service pipe, depending on method and ground conditions. Partial lead pipe replacement will not solve the problem in many cases and can even make matters worse, at least in the short term, and attempts at total lead pipe replacement that include internal domestic lead pipes have failed because of a lack of cooperation from home-owners.

Elevating the pH of water supplies to between 8 and 10.5 will much reduce plumbosolvency, but may not be sufficient for meeting modern day standards for lead in drinking water. Ortho-phosphate is the most effective corrosion inhibitor for reducing plumbosolvency, whereas poly-phosphates can exacerbate plumbosolvency. Ortho-phosphate doses are water supply specific and normally range from 0.5 to 2.0 mg/l (P) to achieve effective plumbosolvency control, 99% compliance being possible with the WHO Guideline Value. Such doses of ortho-phosphate are considered to be entirely safe, given that ortho-phosphate concentrations are many thousands of times higher in many carbonated soft drinks. In general, the environmental impact of dosing ortho-phosphate to a water supply system will be small, although specific local circumstances should be considered. Ortho-phosphate dosing will avoid major disruptions in Cities and Towns. Dosing ortho-phosphate to control plumbosolvency costs considerable less than lead pipe replacement, with a typical unit cost in the UK (in 2009) of 0.5 cents (€) per cubic metre.

Laboratory testing at 25°C using sections of new lead pipe is reproducible and can determine the plumbosolvency of drinking water and test results are available within one month. The impact of pH changes on plumbosolvency can be quickly investigated and the test can determine which corrosion inhibitor works best and its dose response. Computational modelling can predict lead emissions across an entire water supply zone, The modelling system has been validated in UK case studies and the model can predict the results of dosing a corrosion inhibitor and assist in dose optimisation and predict the results of partial lead pipe replacement.

1 Sources of lead in drinking water

1.1 Water resources

With rare exception all source waters have relatively low lead. Ground-waters generally have very low lead and river waters sometimes contain detectable lead from industrial discharges or acidic mine drainage. However, even in these cases, much of the lead is removed by water treatment.

1.2 Distribution systems

The water mains used to distribute drinking water have normally been constructed from cast-iron, ductile iron, asbestos cement and, more recently, from plastics (for example, MDPE: medium density polyethylene). Leaching of lead from these materials does not normally occur. Very exceptionally, lead water mains have been used to distribute drinking water. The only known occurrences are in Ireland where their replacement has high priority (Devaney, 2009).

1.3 Lead pipe service connections

The connection between the water main and the buildings where drinking water is consumed (houses, apartments, institutional buildings and industrial premises) was commonly made using a lead, lead-lined or lead-alloyed service pipe in many countries up until the early 1980s. Lead pipes were used because of their resistance to corrosion (compared to iron) and their malleable nature which minimised fracture and leakage under changing ground conditions.

Lead service pipes often supply a single building, in which case the internal pipe diameter is commonly 12 mm. Larger pipe diameters, up to 38 mm, are used for connections to multiple dwellings such as apartment blocks. Intermediate diameters are used for shared connections to a small group of dwellings. The length of the lead service pipe is dictated by the location of the water main and the depth of property frontage. The typical length in an urban/suburban environment will be between 5 and 100 metres. In rural environments, lengths up to 300 metres have been encountered.

Lead pipe "pig-tails" and "goose-necks" were also used to join other pipe materials, but the lengths involved should be limited to a few metres.

1.4 Domestic lead pipe-work

Lead pipes were also used extensively within buildings to convey drinking water to points of use. However, it is likely that their use was largely superseded by copper piping from the 1950s. In some countries the lead pipes within domestic dwellings have been gradually replaced by copper or plastic during refurbishment, particularly in the modernisation of kitchens. However, in some countries it was common practice to bury internal pipes within walls and their replacement has occurred much less. The variable configuration of pipe-work at a single dwelling is illustrated by Figure 1.1.

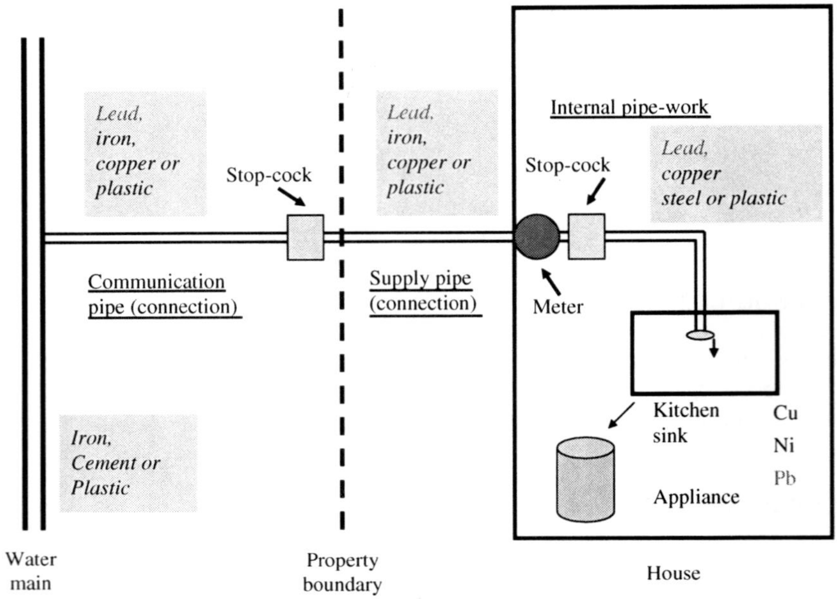

Figure 1.1 Pipe-work configuration at a single dwelling

Taking lead service pipes and internal lead pipe-work together, the extent of occurrence varies nationally, in Europe, from <5 to 50% of domestic dwellings (see Section 5.2) and it is possible that in the older districts of many Cities and Towns, the percentage of dwellings with a lead pipe might be much higher, possibly up to around 90%. "Legacy" lead can also be present in non-lead domestic or institutional pipework (such as old galvanized steel or copper). Lead can accumulate in the pipe scales or pipe-work sediments for decades of exposure to upstream lead sources, even those releasing relatively low concentrations of lead into the water.

1.5 Brass fittings containing lead

Brass fittings have been used very commonly in conjunction with copper pipe-work, as elbows, connectors and valves. Brass manifolds have also been used to distribute drinking water to a group of dwellings from a single mains connection. Water meters, pressure control, and flow control devices have historically also been commonly made from leaded brasses. In the US, there is a relatively small number of brass service lines reported, primarily in the northeast and midwest areas. They were apparently mostly installed in a block of time over the later 1920's, though records are very poor about this.

Brass is an alloy containing copper, zinc and lead, the latter component conferring better machining qualities. Brass is considered to be "lead-free" in the US if the lead content is <8%. New brass alloys with minimal lead contents, or those certified against recent stringent leaching standards in the US and Canada, should not pose problems with lead leaching. However, there are recent examples of very high lead contents (80%) at the internal surface of brass manifolds, albeit with a general lead content of around 7%, presumed to be due to poor moulding. In this situation, lead leached to 800 μg/l after 8 hours water contact and 1,600 μg/l after 16 hours water contact (Hepple, 2008).

1.6 Galvanic corrosion of solders containing lead

Solder has also been used very commonly in the jointing of copper pipe-work. If the solder contains lead and if the solder is exposed to water flow at the joint, sacrificial corrosion of lead can occur as a consequence of galvanic (electro-chemical) effects. The use of lead-containing solders was banned in Europe and North America from the early to mid-1980s.

Galvanic corrosion can cause high, and often erratic, concentrations of lead to leach into drinking water. Lead leaching even from very old solder that has been well-passivated, can spike dramatically in response to certain changes in water treatment that affect galvanic corrosion. These changes generally involve treatments that increase the chloride to sulphate mass ratio of the water (Edwards and Triantafyllidou, 2007) including changes in coagulant from ferric sulphate or alum to ferric chloride or polyaluminum chloride (Renner, 2006), chloride based anion exchange, and desalination. In rare cases the lead can remain high for the long term.

Low pH and aggressive anion gradients in the proximity of wrapped solder joint galvanic connections can also locally mobilize high concentrations of dissolved and particulate lead.

1.7 Plasticizers

Historically, some plasticizers associated with plastic pipes contained lead and leaching of lead into drinking water was found to be possible. Such plasticizers are not now used and while the problem is detectable it is not associated with high level lead leaching.

1.8 Soluble and particulate lead fractions

The principal mechanism for lead leaching from lead pipes to drinking water is the dissolution (or sloughing) of the corrosion film that forms on the inside surface of the pipes. The most common initial oxidation product of lead metal, divalent lead oxide, rapidly converts to lead(II) carbonate or lead(II) hydroxycarbonate compounds in the presence of carbonate or bicarbonate ions in the water. Divalent lead carbonate and hydroxycarbonate solid phases have low solubility, dependent upon the pH, alkalinity and temperature of the water. However, equilibrium solubilities may easily exceed 0.1 mg/L at even circumnerutral to slightly basic pH, and thus greatly exceed modern day drinking water lead standards. Figure 1.2 illustrates the importance of pH in determining lead concentrations, particularly for low alkalinity waters. The interrelationships amongst pH, alkalinity, dissolved inorganic carbon (DIC), and orthophosphate with respect to plumbosolvency have been extensively studied and reviewed, and the results have been shown to be highly reliable in predicting treatment targets and strategies for plumbosolvency in the pH range of approximately 7 to 9.5 (Schock, 1989; AWWARF, 1990; Schock *et al.*, 1996; Schock and Lytle, 2010). However, it should be appreciated that pH elevation alone will often not meet the modern-day standards for lead in drinking water in many water supply zones, particularly if high numbers of lead service lines are present.

In some water supply areas, problems are experienced with the corrosion of old cast-iron water mains, such that iron (red-water) discolouration can become an aesthetic problem with iron concentrations exceeding several parts per million (mg/l). In such cases, the loose iron corrosion deposits can settle within a lead pipe and absorb lead; it is likely that this absorption enhances lead dissolution from the lead corrosion deposit as the equilibrium concentration for the dissolved lead is given less opportunity to be realised. Any disturbance of the loose deposits, such as the scouring effect of high flow, can cause elevated concentrations of lead in the drinking water. As an approximation, lead concentrations can double as a consequence of the interaction with loose iron deposits. Particulate lead may also arise from the physical sheer of pieces of the lead corrosion deposit from within the lead pipe, as a consequence of physical damage (as can occur in partial lead pipe replacement). Vibration from heavy road traffic might also cause pieces of the lead corrosion deposit to sheer. There is some evidence (Cardew, 2009) that the lead

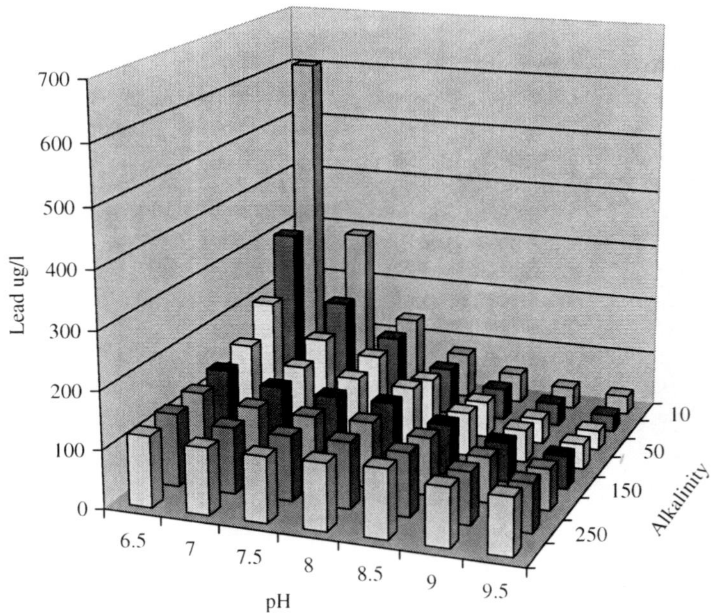

Figure 1.2 Equilibrium concentration of lead in drinking water as a function of pH and alkalinity (mg/l as $CaCO_3$) – from Croll (2000)

corrosion deposits are less stable physically in areas with low alkalinity and organically coloured waters, and that ortho-phosphate dosing has improved stability with associated less occurrence of particulate lead.

There is strong evidence that natural organic matter, particularly the humic and fulvic acids associated with colour, increases the dissolution of lead from lead pipes (Cardew, 2009) by up to ten times (Hayes and Skubala, 2009a). This can be explained by chelation of the lead by the poly-anionic organic matter. In consequence, ortho-phosphate doses must be higher for reducing lead dissolution in waters which have an appreciable organic content (>3 mg/l TOC). Organics from sewage effluent and algae can also increase lead dissolution.

Key points:

1. *The main source of lead in drinking water is from lead service pipes and the lead pipes inside dwellings.*
2. *In the older districts of some Cities and Towns, up to 90% houses can have a lead pipe.*
3. *Occasionally, significant problems can be caused by lead leaching from brass and from lead-containing solder.*
4. *Lead dissolution to low alkalinity waters can be significantly reduced by pH elevation, even with alkalinity adjustment, but may not be enough on its own to meet the modern-day standards for lead in drinking water.*
5. *Iron "red-water" discolouration will exacerbate lead dissolution from lead pipes and may need to be minimised if fully effective plumbosolvency control is to be achieved by corrective water treatment.*
6. *Organic matter, particularly humic and fulvic acids, will exacerbate lead dissolution from lead pipes and will need to be minimised if fully effective plumbosolvency control is to be achieved by corrective water treatment.*
7. *Poly-phosphate dosing can also exacerbate lead dissolution from lead pipes, and is not a preferred treatment strategy for plumbosolvency.*

2 Regulatory background

2.1 World Health Organization guidelines for drinking water

The third edition of the World Health Organization's (WHO, 2004) Guidelines for Drinking Water (and subsequent revisions) includes a guideline for lead and recommendations for drinking water safety planning, both of relevance to this Best Practice Guide (the WHO guidelines also extend to many other aspects of drinking water). Drinking water safety planning is strongly advocated in the Bonn Charter (IWA, 2004).

WHO guideline for lead in drinking water

The toxicity of lead is well established and the WHO has established a provisional tolerable weekly intake of 25 μg/kg body weight (equivalent to 3.5 μg/kg body weight/day). Using the weight of an infant of 5 kg, a consumption of drinking water of 0.75 litres/day and an exposure contribution of 50% from drinking water, WHO has established a guideline value of 10 μg/l for lead in drinking water (WHO, 2004), as a weekly average concentration.

The way in which this guideline value should be implemented is not specified. However, it is generally assumed that this guideline value applies at individual dwellings and to their individual occupants, as it is aimed at health protection.

The Bonn Charter

The Bonn Charter for Safe Drinking Water was published in September 2004 by the International Water Association. Its goal is "good safe drinking water that has the trust of consumers". It is aimed at all stakeholders from governments to consumers and sets out the principles of an effective framework for managing drinking water quality and the responsibilities of key parties.

Its key principles are:

(1) Management of the water supply chain must be holistic.
(2) Systems to ensure drinking water quality should not only be based on end-of-pipe verification and should incorporate risk assessment and risk control.
(3) Close cooperation is required between all stakeholders.
(4) Communication between stakeholders should be open, transparent and honest.
(5) Roles and responsibilities must be clearly defined.
(6) Decisions about standards should be transparent.
(7) Water should be safe, reliable and aesthetically acceptable (albeit standards may vary regionally and over time).
(8) The price of water must not be prohibitive in meeting fundamental domestic needs.
(9) Assurance methods should be based on "best science" and be sufficiently flexible to meet different regional situations.

© 2010 The Authors. *Best Practice Guide on the Control of Lead in Drinking Water*. Edited by Dr Colin Hayes. ISBN: 9781843393696. Published by IWA Publishing, London, UK.

The Charter proposes a framework for the delivery of safe and reliable drinking water, incorporating the development of water safety plans and the measurement of drinking water quality against relevant standards. The framework is illustrated in Figure 2.1.

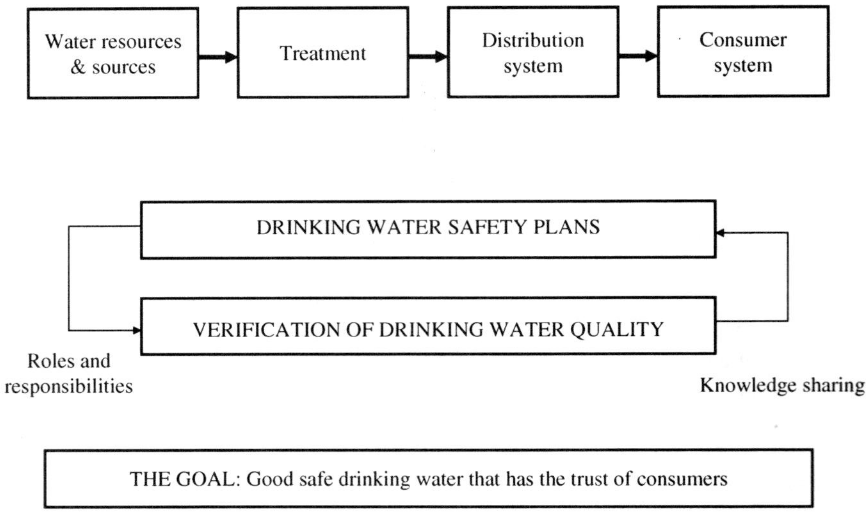

Figure 2.1 Water safety framework

The Charter specifically draws attention to its support of water safety plans as described by the WHO in their 3rd Edition of Drinking Water Quality Guidelines and will provide a foundation for significant improvement in water supply, worldwide. It reinforces the view that effective plumbosolvency control must be holistic.

Drinking water safety planning

The WHO Guidelines (2004) devote an entire chapter to the topic of drinking water safety planning. It is recommended that a risk assessment and risk management approach should be implemented in the design and operation of water supply systems, additional to the verification of water safety by sampling. The risk assessment and risk management approach should extend from "source to tap", that is, the entire water supply chain. This is considered further in Chapter 6 in the context of risk assessment in plumbosolvency control.

The key steps in developing a water safety plan are:

(1) Assemble the development team.
(2) Document and describe the water supply system.
(3) Undertake a hazard assessment and risk characterization.
(4) Prepare a flow diagram to illustrate the system and the risks identified.
(5) Identify control measures.
(6) Define how the control measures will be monitored.
(7) Establish verification procedures.
(8) Develop supporting programmes (e.g.: training, upgrade and improvement, R&D).
(9) Prepare management procedures.
(10) Establish document control and communication procedures.

In the context of plumbosolvency control, drinking water safety planning should focus on:

(1) The extent of occurrence of lead pipes in the water supply system.
(2) The corrosivity (plumbosolvency) of the water supplies to lead, as influenced by pH and alkalinity.
(3) The extent of exacerbation of plumbosolvency by natural organic matter.
(4) The extent of iron discolouration and its potential for particulate lead problems.
(5) The adequacy of controls over brass fittings and solder.

The elements of the water safety plan should also focus on the issues related to lead pipe replacement, including particulate lead, and the potential galvanic effects of partial lead service replacement.

2.2 EU drinking water directives

Directive 80/778/EC

The first EU Drinking Water Directive (80/778/EC) set a standard for lead in drinking water of 50 μg/l, qualified by the words "in running water". Most Member States interpreted this as meaning that either the standard applied to the water flowing through the distribution network (before contact with lead pipes) or it applied to samples taken after flushing from consumer taps. Unsurprisingly, very few or no problems with lead in drinking water were identified in these Member States.

The two principal exceptions were: (i) the UK, which implemented the standard on the basis of random sampling (see Chapter 3) and found that corrective actions were required in some areas, particularly in areas with low alkalinity water, and (ii) the Netherlands, where wide-spread reductions in plumbosolvency were initiated by pH elevation and centralised water softening.

Directive 98/83/EU

Presently, the Member States of the European Union have to comply with the second Drinking Water Directive (98/83/EU), which became a legal requirement from December 2003 (unless a derogation applies to a new Member State). It sets an interim standard for lead in drinking water of 25 μg/l and a standard of 10 μg/l that becomes a legal requirement from December 2013. Both standards relate to the average weekly concentration of lead ingested by consumers, the same basis as the WHO guideline value.

The Directive requires:

- Article 6(1): that the parametric values set for lead shall be complied with, in the case of water supplied from a distribution network, at the point, within premises or an establishment, at which the drinking water emerges from the taps that are normally used for human consumption.
- Article 7(1): that samples should be taken so that they are representative of the quality of the water consumed throughout the year.
- Annex I, Note 3 of Part B: that the parametric value applies to a sample of water intended for human consumption obtained by an adequate sampling method at the tap and taken so as to be representative of a weekly average value ingested by consumers; where appropriate the sampling and monitoring methods must be applied in a harmonised fashion to be drawn up in accordance with Article 7(4); and, Member States must take account of the occurrence of peak levels that may cause adverse effects on human health.
- Annex 1, Note 4 of Part B: that Member States must ensure that all appropriate measures are taken to reduce the concentration of lead in water intended for human consumption as much as possible during the period needed to achieve compliance with the parametric value.

According to Article 6(2), non-compliance due to the domestic distribution system in private premises is not the responsibility of the Member States. However, Article 6(3) requires that in relation to these circumstances, when there is a risk of non-compliance, Member States shall none-the-less ensure that (i) appropriate measures are taken to reduce or eliminate the risk, such as advising property owners of any possible remedial action they could take, and/or (ii) other measures, such as appropriate treatment techniques, are taken to change the nature or properties of the water before it is supplied so as to reduce or eliminate the risk of non-compliance.

It should be noted here that the Member States failed to agree a harmonised monitoring method. In consequence, some EU countries have either not sampled for lead in drinking water at all or have used inappropriate sampling methods. Further, the vague requirements of Article 6(3) have largely been overlooked, as have the requirements of Note 4 to Part B of Annex 1.

Revision of Directive 98/83/EU

An *ad-hoc* Working Group has recently put forward recommendations to the European Commission for revision of the Directive in relation to sampling and monitoring (Hoekstra *et al.*, 2008). In summary, the Working Group has recommended (as relates to plumbosolvency control) that:

(1) Risk assessment and risk management strategies, including improvement plans, are put in place in the management of water supply systems.
(2) Operational monitoring, additional to compliance monitoring, is undertaken in order to properly establish risks.
(3) Zonal assessment monitoring is based on random daytime sampling.
(4) Attention must be given to metal leaching from domestic pipe-work systems, in particular to lead.

Assuming that these recommendations are adopted by the Member States, there will be a clear requirement for problems with lead in drinking water to be quantified and for improvement measures to be taken to protect public health, where necessary. This will be a very significant step forward, bearing in mind the extent of current difficulties in the implementation of standards for lead in drinking water. The incorporation of risk management strategies is particularly important as it will require Member States and their water suppliers to proactively determine the extent of plumbosolvency in their areas and to take appropriate improvement action.

Issues that will need to be resolved are:

(1) The relationship between assessed risks from lead in drinking water, risk reduction programmes and compliance with the lead standards of the Directive.
(2) The distribution of legal responsibilities between water suppliers (lead service lines) and householders (internal plumbing) and how these will relate to risk reduction programmes.

2.3 US Lead Copper Rule

The Lead Copper Rule (LCR) was issued in 1991 as part of the implementation of the Safe Drinking Water Act 1986 (SDWA). The LCR is implemented at State level, mainly through approval of State regulations approved by the responsible USEPA Regional Office for that jurisdiction.

The LCR sets a Maximum Contaminant Level Goal of zero for lead and an Action Level (AL) for lead of 15 μg/l. Compliance with the AL for lead is based on the 90th percentile concentration determined from sampling of homes identified to be at highest risk of elevated lead based on documented occurrence of lead service lines and leaded solder. A one litre, first draw sample is collected from the cold water (usually kitchen) tap considered to be the principal source of water for human consumption. It is collected from the selected homes after a minimum of 6 hours of standing (stagnation). Surveys must be undertaken twice every six months from designated water supply areas, the number of samples in a survey ranging from 5 to 100 for standard monitoring and 5 to 50 for reduced monitoring, for populations ranging from <100 to $>100{,}000$, respectively.

Tiers of risk level for high lead must be inventoried for homes at each utility to determine sampling sites. These are as follows:

- Tier 1: single family residential houses with a lead service line or with lead soldered plumbing installed after 1982.
- Tier 2: multiple family residential houses or other building types with the same plumbing characteristics as Tier 1, but only if an insufficient number of Tier 1 houses are available.
- Tier 3: single family residences with lead soldered copper plumbing installed before 1983, but only if an insufficient number of Tier 1 and 2 houses/buildings are available.

LCR samples must be conducted in homes of the lowest possible tier. That is, tier 2 homes are sampled only if a sufficient number of Tier 1 homes are not available, and Tier 3 homes are sampled only if a sufficient number of Tier 1 and 2 homes are not available.

If a water supply area fails to comply with the AL for lead then optimal corrosion control treatment must be determined and then implemented, and a programme of lead service line replacement for pipe segments owned by the public water system must be subsequently initiated at a minimum rate of 7% per annum.

For large water systems, the LCR requires that the 90th percentile lead level must reflect 'optimal corrosion control treatment,' which is defined under the regulation as the lowest possible lead levels without violating other SDWA primary regulations. The frequency of exceeding the 0.015 mg/L 90th percentile Action Level will trigger other activities, such as public notification and partial lead service line replacement.

There are a range of issues relating to the effectiveness of the LCR for lead:

(1) Detailed interpretations vary between the States, particularly on what is considered to be optimum corrosion control.
(2) The criteria for selecting sampling locations enable houses without lead service lines to be included in the surveys that are undertaken, making it easier for compliance with the AL for lead to be achieved (in this context, it is likely that the criteria based on lead solder are less relevant).
(3) The first draw one litre samples taken will comprise or include any water stood in non-lead pipework, in-situ between a lead service line and the tap outlet, effectively diluting any lead in the water that has stood in the lead service line, again making it easier for compliance with the AL for lead to be achieved (this aspect is considered in more detail in Chapter 3).
(4) Recurring access to the same houses for sampling after 6+ hours of stagnation is difficult to achieve both for each monitoring period, and for continuation of the monitoring site pool over time. It is also not amenable to direct sampling by water utility staff, or their contractors.
(5) Requirements for partial replacement of utility-owned lead pipes may actually aggravate consumer lead exposure in some cases.
(6) Some water systems have compromised optimal plumbosolvency and lead release control because of difficulties in meeting other drinking water regulatory standards.

Presently, the Lead and Copper Rule is undergoing review, and changes in requirements may emerge at the end of the process.

2.4 Developing legislation in Canada

The Guidelines for Canadian Drinking Water Quality, published by Health Canada, state a Maximum Allowable Concentration of 0.010 mg/l (10 µg/l) and is based on a sample taken after flushing. The Guideline was developed in 1992 and subsequently implemented by the provinces and territories and remains the health-based lead limit for drinking water in Canada.

However, Health Canada (2009) has since developed guidance on controlling corrosion in distribution systems which complements the health-based Guideline. This guidance outlines sampling methodology and actions to undertake to reduce potential exposure to lead from drinking water resulting from corrosion.

The guidance provides two options for residential sampling. The first option is intended for systems in which lead service lines are present and where a minimum of 50% of the sampling sites target lead service line residences. It is based on:

(1) A first tier, similar to the US LCR.
(2) A second tier where utilities conduct additional sampling at 10% of the sites sampled in Tier 1 with the highest lead concentrations (above 0.015 mg/l). This sampling is intended to provide the profile of the home plumbing system and determine the source of lead. This helps to better inform approaches to consider for corrosion control.

The second option is intended for systems in which only residences with lead service lines are sampled. It is based on:

(1) A first tier standard of 0.010 mg/l from the average of 4 sequential 1 litre samples for each site, assessed by the 90th percentile concentration of samples taken after a minimum of 30 minute standing. This sampling is intended to provide the profile for the home plumbing system and determine the source of lead.
(2) Conduct additional sampling at 10% of the sites sampled in Tier 1 with the highest average lead concentrations (above 0.010 mg/l). This sampling when used before and after corrosion control, can better evaluate the efficacy of the measures.

This new approach continues to be studied in Canada although the Province of Ontario has implemented a regulatory standard of 10 μg/l for a 30 minutes stagnation sample.

For Ontario utilities, compliance is based on the 90th percentile of the greater of two sequential 1 litre samples after the water has stood for 30–35 minutes (Ontario Ministry of the Environment, 2007).

The issues raised with the US LCR will also apply in Canada, if the approach is implemented. Dilution effects can also affect the results of 30 minutes stagnation sampling, as discussed in more detail in Chapter 3. There is also the issue of non-equilibrium concentrations which may not be reflective of peak values.

2.5 Protocol on Water and Health

The Protocol on Water and Health sets a number of legal obligations for its 24 participating States (United Nations Economic Commission for Europe/WHO, 2007) in relation to the prevention and control of "water-related disease". Article 2(1) defines "water-related disease" as "any significant adverse effects on human health, such as death, disability, illness or disorders, caused directly or indirectly by the condition, or changes in the quantity or quality, of any waters".

The health impact of lead in drinking water clearly falls within this definition and the Protocol requires:

- adequate supplies of wholesome drinking water – Article 4(2)(a);
- effective systems for monitoring situations likely to result in water-related disease – Article 4(2)(e);
- preventative action to avoid incidents of water-related disease, with special consideration for vulnerable people – Articles 5(e) and 5(k);
- establishing and publishing local targets, that need to be achieved or maintained for a high level of protection against water-related disease – Articles 6(2) and 6(3);
- participating States shall establish surveillance and early warning systems, contingency plans and response capacities – Article 8(3);

- the development of indicators to show how far action on water-related disease has been successful – Article 12(b);
- preparation of water-management plans and schemes for improving water supply – Article 14(a).

The Protocol's Working Group on Water and Health, at its meeting in Geneva in July 2009, adopted proposals to include lead in drinking water as a core target health parameter, as a common indicator across the 24 Parties to the Protocol. Initial benchmarking and improvement planning processes were thereby initiated. It can be noted that 15 of the Parties are EU Member States and will therefore be tackling the issue of lead in drinking water ahead of the proposed revision of the Drinking Water Directive.

Guidelines for benchmarking problems with lead in drinking water have been prepared by COST Action 637 in liaison with the EC Joint Research Centre and the World Health Organization (Hoekstra *et al.*, 2009). These guidelines are based on risk assessment to identify a representative number of water supply areas that are then surveyed by random daytime sampling to determine priorities for corrective action (see Chapters 3 and 6).

Key points:

1. *The regulatory approaches in Europe and North America have been beset by problems due to difficulties associated with sampling for lead in drinking water.*
2. *In Europe, there are proposals to normalise compliance monitoring by the use of random daytime sampling.*
3. *A trend is emerging that will include risk assessment and risk management in the regulation of lead in drinking water.*
4. *In consequence, water suppliers will in future need to look more deeply into the causes and extent of plumbosolvency problems in their area.*
5. *It is likely that optimisation of corrosion control will require further attention.*

3 Sampling and monitoring

3.1 Role of monitoring in plumbosolvency control

Compliance monitoring

The regulations that establish standards for lead in drinking water require compliance to be demonstrated by monitoring. The definition of ***compliance monitoring*** is the monitoring at the point of compliance to verify that water supplied for human consumption is in compliance with its quality requirements.

In the US, this is achieved by periodic surveys at consumers' taps with sample numbers reflecting the population of the water supply system being assessed (Table 3.1). Whether the system complies or fails the LCR is clearly defined on the basis of observed 90th percentile concentrations. Water quality monitoring points must also include both the entry points to the distribution system, as well as locations in the distribution network itself.

Table 3.1 Sample numbers for lead required by US LCR

Population served	Required number of sampling sites: standard monitoring	Required number of sampling sites: reduced monitoring
>100,000	100	50
50,001 to 100,000	60	30
10,001 to 50,000	60	30
3,301 to 10,000	40	20
501 to 3,300	20	10
101 to 500	10	5
<100	5	5

Standard monitoring requires two surveys every six months. This can be relaxed to reduced monitoring: (i) annual survey if two consequetive 6-monthly surveys comply; (ii) triennial survey if the 90th percentile concentration is less than 5 μg/l in two consequetive 6-monthly surveys, and (iii) survey every 9 years if the population served is <3,300 and the 90th percentile concentration is less than 5 μg/l in two consequetive 6-monthly surveys and the system is free of lead pipes and leaded brass and solder.

Related water quality parameters (pH, alkalinity, calcium if carbonate stabilization is used, and orthophosphate or silicate if corrosion inhibitors are dosed) must also be assessed but at a lower frequency.

In the EU, a minimum number of samples must be taken each year from consumers' taps, again dependent on the size of the water supply system being assessed. The frequencies for compliance monitoring for lead are given in Table 3.2.

Table 3.2 Minimum compliance monitoring for lead in the EU

Volume of water distributed[a]	Number of compliance samples[b]
m^3/day	n/year
$\leq$10	[c]
$>$10–$\leq$100	[c]
$>$100–$\leq$1,000	1
$>$1,000–$\leq$10,000	1 +1 for each 3,300 m^3/day and part thereof of the total volume
$>$10,000–$\leq$100,000	3 +1 for each 10,000 m^3/day and part thereof of the total volume
$>$100,000	10 +1 for each 25,000 m^3/day and part thereof of the total volume

[a] Minimum frequencies for compliance monitoring should depend on the risk assessment made in the water risk management strategy and could be higher.
[b] As far as possible, the number of samples should be distributed equally in time and location.
[c] The frequency is to be decided by the Member State concerned.

It is recognised (Hoekstra *et al.*, 2008) that these minimum frequencies (Table 3.2) are inadequate for operational control purposes.

Inventory monitoring

The definition of ***inventory monitoring*** is the monitoring to estimate the lead problem in a water supply zone or country. To support the implementation of the Protocol on Water and Health, the guidelines for the sampling and monitoring of lead in drinking water (Hoekstra *et al.*, 2009) recommend surveys based on random daytime sampling with the sample numbers in each water supply system surveyed as shown in Table 3.3.

Table 3.3 Inventory monitoring based on random daytime sampling

Production volume in water supply zone (m^3/day)	Total number of samples in each survey	Number of samples per month over a 6 month period
$<$10,000	180	30
10,000 to $<$15,000	240	40
$>$15,000	300	50

It is recommended that the appropriate number of samples should be taken evenly over a period such that seasonal factors are taken into account. Normally the minimum sampling period should be six months that spans equally both higher and lower water temperatures. This is because lead dissolution into water will have approximately twice the concentration in the summer than in the winter.

Operational monitoring

The definition of ***operational monitoring*** is the monitoring activity to check the quality of source water, and to validate the operation of the water treatment plant, the distribution network and the domestic distribution system up to the tap.

The key parameters in operational monitoring for plumbosolvency control are:

pH: (i) at the water treatment plant to check process control, and (ii) through the distribution network to check variation caused by mixing of source waters and to check water stability (pH reductions can occur with poorly buffered low alkalinity waters due to the microbiological activity of biofilms, particularly in corroded cast iron water mains, and pick-up of carbon dioxide at storage points). Evaluation should be undertaken of pH reduction related to nitrification in the distribution system when chloramines are used. Although this is less common in the EU, chloramination is becoming more common in Canada and the US.

Alkalinity: (i) at water treatment plants that feed systems requiring lime treatment to improve buffering capacity, and (ii) through the distribution network to check variation caused by mixing of source waters.

Ortho-phosphate when dosed for corrosion control: (i) at the water treatment plant to check process control, and (ii) through the distribution network to check variation caused by mixing of source waters and to check water stability (phosphate attenuation should not normally exceed 10%) – random daytime sampling from consumers' taps will provide the most representative check of variation across the water supply system.

Lead: several approaches can be used in combination to demonstrate the success (or otherwise) or corrective actions: (i) benchmark monitoring at selected houses which have lead pipe-work (see Section 3.4), (ii) lead pipe test rigs at strategic locations (see Section 10.2), and (iii) zonal surveys – random daytime sampling from consumers' taps will provide the most representative check of variation across the water supply system.

It is common for both pH and ortho-phosphate monitoring at water treatment plants to be undertaken by on-line monitors, often linked via telemetry to a central control facility.

Definition of water supply systems for monitoring and control purposes

In the simplest case, the water supply system will have its own unique source and treatment plant and the water quality across the system will be fairly uniform. Corrective treatment for reducing plumbosolvency should also have a fairly uniform effect. Unless iron corrosion problems are significant, attenuation of ortho-phosphate within the distribution network will be slight and should not exceed a 10% loss at consumers' taps. Multiple sources to a water supply system may differ markedly in their water quality characteristics, such as the augmentation of a river derived source by local ground-water abstraction. In such cases, corrective treatment to reduce plumbosolvency must take account of the different treatment needs of each source water and endeavour to satisfy the treatment needs of the entire water supply system. This could mean dosing ortho-phosphate at one or more source locations at a level that is higher than required specifically by one or more source waters, in order to ensure that the higher dosing requirement of another source water is not diminished by mixing or blending of the source waters within the distribution network. Integated water supply systems that are fed by numerous source waters may need to be defined in relation to the feasibility of locations for corrective water treatment. As a general rule, the designation of a water supply system should be determined by reference to its source water(s). However, localised problems may require such designations to be broken down into smaller units that reflect better localised conditions, such as housing age.

A major significance of the designation of a water supply system is that its size, as defined by water consumption or population, will determine the extent of both regulatory and operational monitoring. Whereas minimum compliance monitoring for lead in Europe increases with the size of the water supply system (Table 3.2), the US Lead Copper Rule defines a maximum number of compliance samples, regardless of system size for systems serving more than 100,000 (Table 3.1). In this context, there are no guidelines at present for inventory monitoring other than recommendations in relation to the Protocol on Water and Health (Table 3.3). Operational monitoring to determine plumbosolvency control requirements and demonstrate subsequent improvements should be applied to system sizes no greater than 100,000 population served, if at all possible. This will provide the resolution needed to give confidence in the monitoring results and avoid the over-looking of significant localised problems.

Investigating problems

Consideration should be given to the investigation of any sample result that falls outside pre-determined limits. For lead at consumers' taps, non-compliance with the relevant regulatory standard may reveal atypical circumstances such as recent plumbing alterations. Judgement and pre-agreed protocols will be needed to determine if such atypical results should be included in compliance assessment. It is recognized that in systems with high levels of particulate lead, extraordinary variability in actual concentrations can occur even when proper samples are collected under a rigorous and well-controlled protocol. Problems with high lead concentrations at consumers' taps can be investigated by several sampling methods, to determine legal responsibilities and potential health risks (see Section 3.5).

3.2 Sampling methods

Fully flushed samples

The definition of a ***fully flushed sample*** is a sample that is taken after prolonged flushing of the tap in a premise in such way that stagnation of water in the domestic distribution system does not influence the concentration of lead. In practice a sample is taken after flushing at least three plumbing volumes. In cases where the temperature of the water from the distribution network is cooler than the ambient temperature, an alternative method is monitoring the temperature of the water during flushing until it stabilises.

The main value of a fully flushed sample is to confirm the absence of lead in the distribution system prior to contact with domestic pipe-work. Lead residuals should normally be below 1 μg/l in a fully flushed sample from a consumers' tap. If not, there are a number of possible reasons: (i) flushing has not been sufficient, particularly with very long lead pipes, (ii) there is lead in the distribution network from a lead water main (rare but possible), or (iii) particulate lead problems are being experienced. Fully flushed samples have no role to play in determining compliance with standards at consumers' taps.

First draw samples

The definition of a ***first draw sample*** is a sample that is taken first in the morning before the tap in the premise has been used for other purposes. During the over-night stagnation period no water should be drawn from any outlet within the property (this includes flushing of toilets). If any water is drawn during the stagnation period the result will be invalid. It is common practice for such samples to be taken by consumers and there is only limited control over the quality of the samples. When the sample is taken the tap should be fully opened or as open as possible without losing sample. It is normal for one litre volumes to be sampled at the first draw flow event. The stagnation of water in the domestic distribution system influences the concentration of lead as illustrated by Figure 3.1.

In Figure 3.1, plumbosolvency decreases with curves A to C. The difference between curves A1 and A2 is the shape of the curve, which varies and is water specific.

With 12 mm internal diameter lead pipes, which are often used in the service lines to single family dwellings, the minimum 6 hours stagnation required by the US LCR is fairly close to the time at which maximum solubility equilibrium occurs.

A major weakness in this sampling approach is the dilution of water that has stood in lead pipe-work with water that has stood in non-lead pipe-work between the lead pipe-work and the tap outlet. To illustrate, let us consider a lead pipe joined to a copper pipe up to the tap. A 12 mm internal diameter copper pipe of 8.8 metres length has a volumetric capacity of 1 litre. If the copper pipe between the lead pipe and the tap is 4.4 metres in length, the first draw one litre sample will comprise 50% water that stood in the copper pipe and 50% water that stood in the lead pipe, a 50% dilution effect. If the copper pipe is 8.8 metres in length, or greater, no water that stood in the lead pipe will be captured by the first draw one litre sample. This assumes simple plug flow in the pipes concerned, which is reasonable at the flow rates normally experienced (about 0.1 litre per second, that is 6 litres per minute) as the flow should be turbulent within such small diameter pipes and exhibit plug flow characteristics. Dilution effects can also occur with short pipe lengths when mains water is drawn into the 1 litre sample volume. For the LCR, it is also reasonable to assume that pipe-work will have relatively fresh water in it prior to the stagnation period, as consumers use water for washing, cleaning dishes, flushing the toilet, etc before the embargo on water use begins. In practice, the freshness of the pre-stagnation will vary and it is possible that some water in the non-lead pipe will contain lead.

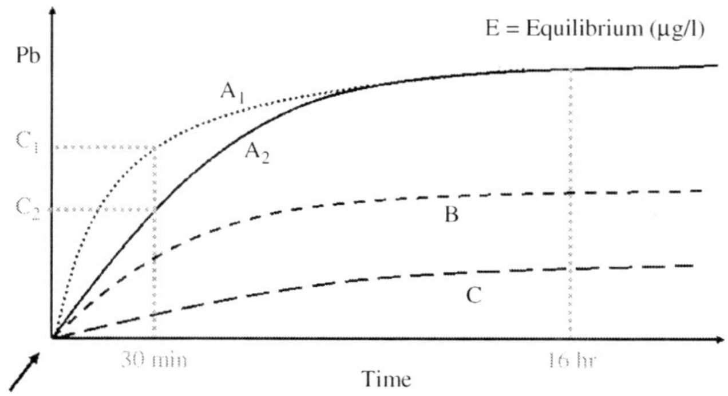

M = Initial mass transfer rate (μg/m^2/sec), the initial slope. C_1 and C_2 are the 30MS concentrations of curves A_1 and A_2. B and C are curves for waters of lower plumbosolvency.

Figure 3.1 Lead dissolution over time (from Hayes, 2009)

Sequential one litre sampling (say three consecutive samples) will minimise the dilution effect but interpretation of the result for each sample will be complicated by the relationship of water drawn from lead pipe-work and non-lead pipe-work as determined by pipe lengths and diameters. Interpretation is further challenged by the amount of lead contained in non-lead pipe-work prior to stagnation.

Proportional samples

The definition of a ***proportional sample*** is a sample that is taken during the use of tap water for human consumption in a property. The sample is an approximation of the concentration of lead that is consumed in that property. In its simplest form of application, this sampling method requires a special consumer-operated device to be fitted to the tap that splits off a small constant proportion of every volume of water drawn for dietetic purposes. There is no control over the quality of the samples taken because it relies on consumer operation. The proportional sample is normally collected during a 1-week period.

This sampling method provides a direct measure of the average lead concentration being emitted at the consumer's tap and will approximate to the lead ingested. The method has a number of constraints: (i) the cost of the sampling device, (ii) the time taken to install and decommission the device, and (iii) it will require a large number of sampling events to characterise the spatial and temporal variation of lead emissions across a water supply system. It should be possible to automate the operation of a proportional sampling device although this will add to the cost constraints in widespread use. The method has potential application for investigating lead emissions at specific locations.

Two studies (van den Hoven *et al.*, 1999; Hayes, 2009) indicate for zonal compliance assessment, that the more logistically favourable random daytime sampling method gives adequately similar results to proportional sampling. For most levels of plumbosolvency, the zonal assessments (i.e.: compliance rates) based on random daytime sampling and proportional sampling are similar, but for a phosphated water with a very low plumbosolvency, random daytime sampling is more stringent.

This equivalence of the two sampling methods does not apply to the assessment of lead emissions at individual properties.

Stagnation samples

The definition of a ***stagnation sample*** is a sample that is taken after prolonged flushing of the tap in a property (see fully flushed sampling) and successive stagnation for a predefined period before the sample is taken from the tap. During the stagnation period no water should be drawn from any outlet within the property (this includes flushing of toilets). If any water is drawn during the stagnation period the result will be invalid. When the sample is taken the tap should be fully opened or as open as possible without losing sample. The stagnation of water in the domestic distribution system influences the concentration of lead in a predefined way, as illustrated by Figure 3.1. A stagnation time of 30 minutes and a 1 litre sample volume are commonly used.

30 minutes stagnation sampling suffers from the same dilution effect as a consequence of water stood in non-lead pipe-work, as described for first draw samples. This effect invalidates 30 minutes stagnation sampling for the purpose of zonal compliance assessment. The analytical error is also larger relative to the observed value itself for the lower concentrations usually encountered in 30 minute stagnation samples. Therefore, the confidence that samples meet the standard is weaker, and there is increased danger that high levels of lead that results for consumers drawing water after longer stagnation, or who get slugs of water directly in longer contact with lead service lines, will be missed.

Sequential 30 minutes stagnation sampling reduces the dilution effect, but interpretation in zonal assessment is complicated by the relationship between pipe lengths and diameters. However, sequential sampling after 30 minutes stagnation can help to identify which part of the pipe-work system, from the connection to the water main to the tap outlet, is responsible for lead emissions, to assist the determination of legal responsibilities.

A further weakness of 30 minutes stagnation sampling is that it does not reflect the average lead concentration emitted from a lead pipe, of relevance to potential health impact, as illustrated by computational modelling data in Table 3.4. This is because the average lead concentration emitted is pipe length dependent whereas 30 minutes stagnation samples are not lead pipe length dependant (assuming no dilution effects from water stood in non-lead pipe-work).

Table 3.4 Predicted average and 30 minutes stagnation sample results from different lengths of a lead pipe[a]

Pipe length (m)	Daily average Pb concentration (μg/l)	30 minutes stagnation Pb concentration (μg/l)
10	4.8	49.6
20	9.6	49.6
30	14.3	49.6
40	18.8	49.6
50	23.1	49.6
100	42.7	49.6

[a] modelled as described in van der Leer *et al.* (2002), assumptions: daily water volume = 450 litres; pipe diameter = 12 mm; water of moderate plumbosolvency (M = 0.1, E = 150); standard daytime water consumption pattern with half hour flow frequency.

Stagnation sampling can be used in the benchmarking of improvement measures based on water treatment, by repetitive sampling at selected properties. The properties must be selected to give an appropriate stagnation sample result that will be able to demonstrate the subsequent changes brought about by the treatment measure. An example is given in Section 3.5.

In Austria and Germany, some water utiliities use 4 hour stagnation sampling to determine weekly average lead concentrations; 2 hour stagnation is also undertaken for this purpose but the results are extrapolated to 4 hours stagnation. These approaches stem from pipe-rig testing protocols developed in Germany but are inconsistent with the 30 minutes stagnation period commonly taken to represent average lead concentrations.

Random daytime samples

The definition of a ***random daytime sample*** is a sample that is taken at a random time of a working day directly from the tap in a property without previous flushing. When the sample is taken the tap should be fully opened or as open as possible without losing sample. The stagnation of water in the domestic distribution system influences the concentration of lead in a random manner. It is common practice to select the sampling points at random and for 1 litre sample volumes to be taken.

When sampling locations are selected at random, the method provides an unbiased assessment of zonal compliance. There should be an appropriate balance between public and private buildings and consumers' taps in individual buildings can be chosen at random from billing lists, electoral registers, post code lists or other similar and suitable lists. A house should not be selected twice in one year. It is common for the sampler to fail to gain access to the property selected for sampling. In such cases, the sampler should simply attempt to gain access to an adjacent property (randomness will be preserved).

Random daytime sampling does not suffer from dilution effects nor from potential problems with selecting representative sampling locations. It is also the most logistically favourable method as the time involved in sampling is very short and it is acceptable for the sampler to optimise the sampling route on each day to optimise travelling. The main weakness of random daytime sampling is that sample numbers need to be sufficiently high to achieve confidence in the assessments made (see Section 3.4). The recommended sample numbers for inventory monitoring (Table 3.3) illustrate this point.

Unlike attempts to monitor at the same homes during each sampling event in the U.S., statistical power in determining trends in lead leaching with time will also be decreased, due to natural variability encountered from home to home. Finally, unless "high risk" sites are represented in the sampling pool, the random program could miss problems arising in pockets of high risk homes. For example, if there were only a few dozen homes known to have lead pipe in a city with hundreds of thousands of homes, a truly random protocol might not sample at any homes with lead pipe. In such cases, supplementary investigations will be warranted.

Random daytime sampling, of metals at consumers' taps, has been a regulatory requirement in the UK since 1989 and the Netherlands since 2004. It is common practice in the UK for such sampling to be undertaken by trained water company personnel, a reflection of stringent regulations that apply.

3.3 Analysis

It is normal practice to analyse for total lead with acid preserved samples using an analytical method capable of achieving a limit of detection of less than or equal to 10% of the relevant standard. Analytical quality control is essential to ensure the integrity of the data obtained.

3.4 Zonal monitoring

Strategy for optimising plumbosolvency control

The purpose of zonal monitoring will be to demonstrate that planned improvements are being achieved. In cases where the selected corrective option is the removal of lead pipes, random daytime sampling alone will be sufficient, if enough samples are taken, throughout the duration of the improvement period. In cases where the selected corrective option is the dosing of corrosion inhibitor (ortho-phosphate), random daytime sampling can be supplemented by stagnation sampling at selected points.

Variation in results from random daytime sampling

The problem of reproducibility in random daytime sampling is illustrated by Table 3.5. Before ortho-phosphate dosing commenced, the zonal failure rates varied fairly substantially, year by year. However, when compared to the UK Government's benchmark for optimising corrective action, that is 5% or more RDT samples exceeding 10 μg/l, the variation in results has no consequence as each year clearly exceeded the benchmark.

Table 3.5 Observed variation in RDT sampling results for a water supply system in the UK

(a) Before ortho-phosphate dosing

Year	N	% >10 μg/l	% >25 μg/l	% >50 μg/l
1990	132	28.0	17.4	9.9
1991	228	24.1	11.0	4.0
1992	61	13.1	6.6	0.0
1993	45	17.8	13.3	2.2
1994	29	10.3	6.9	0.0
1990–4	**495**	**22.4**	**12.1**	**4.6**

(b) After ortho-phosphate dosing

Year	N	% >10 μg/l	% >25 μg/l	% >50 μg/l
1996	29	6.9	0.0	0.0
1997	28	3.6	0.0	0.0
1998	28	3.6	0.0	0.0
1996–8	**85**	**4.7**	**0.0**	**0.0**
1999	9	0.0	0.0	0.0
2000	8	0.0	0.0	0.0
2001	26	3.8	3.8	3.8
2002	28	0.0	0.0	0.0
2003	17	0.0	0.0	0.0
1999–03	**88**	**1.1**	**1.1**	**1.1**

After ortho-phosphate dosing commenced in 1995, a marked improvement was observed over the next three years (1996–8), although dosing was not fully optimised. Thereafter, the full benefit of dosing was demonstrated over the period 1999–03, albeit a single sample exceeded all three lead standards in 2001. Despite this single atypical result, the comparison of the bulked results for 1990–4, 1996–8 and 1999–03 clearly demonstrate the success of ortho-phosphate dosing.

Using statistical tools to assess confidence in RDT sample results

In an attempt to ensure adequate statistical confidence in the RDT sampling used in Scotland to determine the success of corrective action, the Scottish Executive issued guidance on the numbers of samples and permitted numbers of failures (against 10 μg/l) that were required to demonstrate compliance with a 95% target, with at least 95% confidence, as shown in Table 3.6.

Table 3.6 Numbers of RDT samples and permitted numbers of failures in Scotland for 95% confidence in achieving compliance with a 95% compliance target against 10 μg/l (Scottish Executive, 2007)

Number of RDT samples required	Number of permitted failures vs 10 μg/l
59	0
93	1
124	2
153	3
208	5
336	10

Benchmark monitoring at selected properties

As an alternative to random daytime sampling, but preferably as a supplementary monitoring method, stagnation sampling at selected properties can provide useful information on the effectiveness of dosing ortho-phosphate for reducing plumbosolvency. This approach is not relevant if total lead pipe replacement is undertaken as the sole corrective measure.

Properties with a sufficiently long lead pipe must first be selected, on the basis of their suitability to demonstrate changes in plumbosolvency. As a guide, 30 minutes stagnation samples should have a lead concentration of at least 20 μg/l before ortho-phosphate is dosed. Selected properties should then be sampled at least monthly and preferably weekly, until such time as treatment objectives have been achieved. Figure 3.2 illustrates the results from 30 minutes stagnation sampling at a property in the UK, both before and after the commencement of ortho-phosphate dosing (as P). The advantage of this monitoring approach is that it demonstrates the response of in-situ lead pipes.

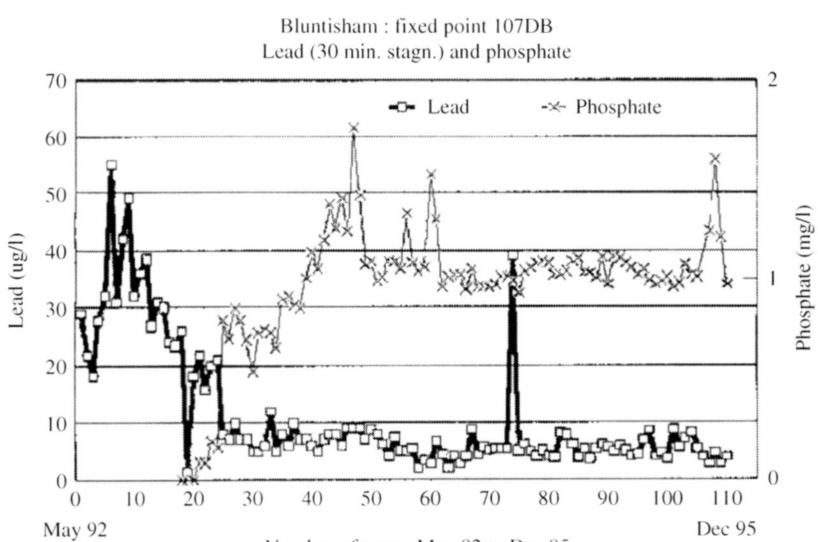

Figure 3.2 30 minutes stagnation sampling at a selected property

In the assessment of plumbosolvency changes in a water supply system, it is advisable to monitor a minimum of 6 suitable properties. The main problem with the approach is that householders tire of repeated sampling visits, move away or die, such that the ability to monitor at an individual property may be lost.

If a water supply system is already being dosed with ortho-phosphate for plumbosolvency control purposes, it will be very difficult to locate suitable properties for benchmark monitoring because lead emissions will be too low to demonstrate change.

The results obtained in a water supply system in Western England over an eight year period are summarised in Table 3.7 for four reference houses.

Table 3.7 Average percentage reductions in 30MS lead concentrations at four reference houses in a water supply system in the UK

Year	Average % reduction	Average ortho-phosphate (μg/l P)
1994	0	<100
1995	5.0	408
1996	17.6	980
1997	33.2	1,060
1998	49.5	1,093
1999	69.9	1,424
2000	78.3	1,402
2001	81.5	1,394

Three points emerge from this fixed point monitoring:

(1) A reduction in plumbosolvency of about 80% was achieved by 1.4 mg/l (P) ortho-phosphate.
(2) Much lower reductions were achieved at doses of ortho-phosphate around 1.0 mg/l (P).
(3) The response to the phosphate dose being applied was about two to three years, much longer than had been anticipated.

The observed lag in responding to ortho-phosphate dosing is particularly important. It is likely explained by the longer time taken for the corrosion deposits within the lead pipes to reach equilibrium with the applied ortho-phosphate dose, due to organic and iron contents. In contrast, the observed lag in responding to ortho-phosphate was between 6 to 12 months in chalk ground-water fed supply systems in Eastern England.

The consequence of these observed lags is that zonal compliance monitoring by random daytime sampling needs to extend to several years to be able to demonstrate a particular ortho-phosphate dose, making dose optimisation very difficult by zonal compliance monitoring alone.

3.5 Investigations at individual properties

An investigation is warranted if a sample from a property contains lead at a concentration above the relevant regulatory limit, and should comprise:

(1) An inspection of the pipe-work from the water main to the kitchen tap to record pipe materials, lengths and diameters (see Figure 1.1).
(2) Enquiries to determine if any recent pipe-work changes have ocurred.

(3) Observation of proximity to heavy road traffic and possible vibration.
(4) Assessment of general water quality characteristics in the area.
(5) Occurrence of particulate lead in samples.

Taking repeat random daytime samples, to attempt to confirm the first sample result, is not recommended, because lead emissions from taps vary too much.

Stagnation sampling will provide a better measure of the extent of lead emission problems, if dilution effects are excluded, within the limitations illustrated by Table 3.4. Sequential stagnation sampling can help to identify which part of the total pipe-work is responsible. Proportional sampling offers the most direct approach for quantifying average lead concentrations, despite logistic and quality assurance constraints.

Key points:

1. *Random daytime (RDT) sampling offers an unbiased approach for zonal compliance assessment but suffers from poor reproducibility, particularly with small sample numbers. If small numbers of homes are at high risk, random sampling can also miss the worst of the lead problems.*
2. *Bulking RDT data over several years enables treatment changes to be adequately demonstrated, albeit allowance must be given to the response time of lead pipes.*
3. *Statistical tools can be used to determine levels of confidence in RDT sampling data.*
4. *Random stagnation sampling is not suitable for zonal compliance assessment because of dilution effects.*
5. *Stagnation sampling at selected properties, for the purpose of zonal compliance assessment, is sensitive to property selection.*
6. *Stagnation sampling can benchmark treatment changes at suitably selected houses and provide a direct measure of the reductions in plumbosolvency achieved.*

4 Health perspectives

4.1 Introduction

There is no doubting that lead is toxic to humans. However, the significance of lead ingestion from drinking water, both in relation to other possible sources (notably from old paint) and to health impacts, is not without contention. Regardless, stringent standards for lead in drinking water apply both in Europe and North America, for the purpose of protecting public health. Control of lead exposure through drinking water should therefore be placed in any over-all scheme of reduction.

The purpose of this chapter is to draw attention to:

(1) The wide range of human toxic effects associated with lead.
(2) Observed relationships between lead in drinking water and lead in blood.
(3) Observed relationships between lead in blood and clinical effects.
(4) Observed relationships between blood lead and reductions in the IQ of infants.
(5) The possibility that lead-induced abortion and lead-induced pre-natal mortalities may warrant attention.

The reader needs to appreciate that there are some inconsistencies between the numerous published health studies and should consult a suitably qualified medical practitioner if more specific health advice is sought.

4.2 Toxicity of lead to humans

Lead is not an essential trace element in any organism and has no known biological function. Its toxicity has been recognised as early as 200BC with both acute and chronic effects arising from lead exposure from various sources (such as air, water and paint). Adverse health effects include: interference with haemoglobin biosynthesis; interference with calcium and vitamin D metabolism; gastrointestinal irritation; dullness; restlessness; irritability; poor attention span; headaches; muscle tremor; abdominal cramps; kidney damage; hallucination; loss of memory; encephalopathy; hearing impairment, gonad dysfunction, and violent behaviour.

Lead in drinking water is considered to be a bio-available source of lead because soluble lead is more readily absorbed in the intestine than lead from dietary sources. However, lead intake is increased from water absorbed into food during cooking. Infants and foetuses are the most susceptible groups to adverse health effects as they absorb 4 to 5 times more lead than adults and consume much more water per unit body weight.

4.3 Lead in drinking water and lead in blood

Studies in Scotland, Wales, Germany and the US have found that high lead concentrations in tap water can correlate with elevated body lead burden. The relationship between water lead and blood lead is illustrated in Table 4.1 from Quinn and Sherlock (1990) and shows that blood lead concentrations of 20 and 25 μg/dl can increase in occurrence with higher average concentrations of lead in water over the range 20 to 100 μg/l.

Table 4.1 Relationship between water lead and blood lead (from Quinn and Sherlock 1990)

Mean water lead concentration (ug/l)	Percentage of children with blood lead concentrations above 20 μg/dl	Percentage of children with blood lead concentrations above 25 μg/dl
100	33	12
50	18	5
30	10	2.3
20	8	1.7

Other studies further illustrate a relationship:

(1) Fertmann *et al.* (2003) correlated blood lead concentrations of 1 to 6 μg/dl with water lead concentrations below 5 μg/l.
(2) Beattie *et al.* (1972) earlier associated blood lead concentrations of 30.9±12.0 μg/dl with water lead concentrations greater than 300 μg/l and associated a blood lead concentration of 19.7 μg/dl with a water lead concentration of 50 μg/l.
(3) Quinn and Sherlock (1990) also drew attention to health studies in Ayr and Glasgow (Scotland) which found a curvi-linear relationship between water lead and blood lead, with an average water lead concentration of 20 μg/l correlating with a blood lead concentration of 10 μg/dl (albeit with much scatter around the best fit lines).

Corrective water treatments of Glasgow's water supplies were applied in 1979 (lime to raise pH) and again in the mid-1980s (additional dosing of ortho-phosphate). Table 4.2 summarises the reductions in water lead and blood lead that were achieved, based on the study by Moore *et al.* (1998). In another study by Watt *et al.* (2000), the geometric mean blood lead concentration in the year 2000 had fallen to 3.7 μg/dl compared to 11.9 μg/dl 12 years previously.

Table 4.2 Reductions in water lead and blood lead in Glasgow (from Moore *et al.*, 1998)

Water lead	Mean blood lead in mothers from Glasgow
83% of RDT samples >50 μg/l in 1977	14.6 μg/dl in 1977
13% of RDT samples >50 μg/l in 1980	8.1 μg/dl in 1980
2% of RDT samples >50 μg/l in 1993	3.5 μg/dl in 1992–93

Blood lead level reductions were also observed in Edinburgh (UK) after drinking water treatment with a 64% reduction in blood lead levels in 8 years between the original and follow-up studies undertaken in 1983–1985 and 1992–1993 respectively (Macintyre *et al.*, 1998).

4.4 Lead in blood and health effects

The action level for blood lead in the United States, used in the prevention of lead poisoning in children, is 10 μg/dl, although it has recently been suggested (Gilbert and Weiss, 2006) that it should be lowered to 2 μg/dl as this lower level can be measured accurately and would encourage further action to reduce childhood lead exposure. However, this may pose a challenge given that the geometric mean blood lead concentration for US children was 1.77 μg/dl for the period of 2003–2004 (CDC, 2005).

The relationship between blood lead burden and adverse health effects is illustrated in Table 4.3 from a range of case studies and related literature sources. Perhaps the greatest health concern associated with lead is reduced IQ in infants:

(1) Tong *et al.* (1996) found a reduction in the mean full scale IQ of 3 points in children aged 11 to 13 years for an increase in lifetime average blood lead concentration from 10 to 20 μg/dl.
(2) Pocock *et al.* (1994) reviewed 26 epidemiological studies from 1979 and found that a doubling of blood lead from 10 to 20 μg/dl was associated with a mean deficit in full scale IQ of around 1 to 2 points.
(3) Canfield *et al.* (2003) found that blood lead was inversely and significantly associated with IQ; the relationship found was non-linear with a decline in IQ of 7.4 points when average lifetime blood lead increased from 1 to 10 μg/dl with a further decline of 4.6 points with each increase of 10 μg/dl in the lifetime average blood lead concentration.
(4) Chiodo *et al.* (2007) determined a relationship ($r = -0.26$, $p > 0.001$) between blood lead and neurobehavioural outcomes in 7 year old children ($N = 506$) over the blood lead range 1 to 20 μg/dl.
(5) Bellinger *et al.* (1992) found a decrease in full scale IQ over a blood lead range of 0 to 25 μg/dl in infants aged 24 months.

Table 4.3 Blood lead burden and reported adverse health effects

Blood lead burden (μg/dl)	Reported adverse health effect	Source
Maternal levels >15 μg/dl	Associated with reduced birth weight	Dietrich *et al.* (1987)
12 to 120 μg/dl	Vitamin D metabolism interference	Mahaffey *et al.* (1982)
Increase from 10 to 20 μg/dl	Reduction in IQ by 3 points	Tong *et al.* (1996)
Increase from 10 to 20 μg/dl	Reduction in IQ by 1 to 2 points	Pocock *et al.* (1994)
Increase from 1 to 10 μg/dl	Reduction in IQ by 7.4 points	Canfield *et al.* (2003)
20 μg/dl	Increased hearing threshold compared to 4 μg/dl	Schwartz and Otto (1987)
Greater than 37 μg/dl	Hypertension	Pocock *et al.* (1984)
40 to 50 μg/dl	Decreased sperm counts	Assennato *et al.* (1987)
40 to 120 μg/dl	A range of neurological and behavioural effects	Agency for Toxic Substances and Disease Registry (2007)

Epidemiological studies undertaken in Glasgow (UK), a city with historically high water lead concentrations, due to highly plumbosolvent waters and the extensive presence of lead plumbing, found evidence of elevated blood lead levels (Moore *et al.*, 1979; Lacey *et al.*, 1985). Studies in Glasgow also suggested development delays and reduced birth weight (Jones, 1989); development deficiencies among children (Beattie *et al.*, 1972); ischemic heart disease, renal damage, gout and hypertension (Moore, 1977) related to high lead concentrations in the tap water.

4.5 Acute affects of lead

The historic account by Troesken (2006), which focuses on the 19th and early 20th Centuries, provides numerous case examples of lead poisoning from drinking water. It also makes reference to the use of lead "plaster" tablets and lead oleate to induce abortion from several medical accounts over the period 1900 to 1905 (Ransom, 1900; Wrangham, 1901; Hall, 1905). Doses of 0.004 grains of lead per day for inducing abortion are quoted which is equivalent to a daily intake of lead of 260 µg or approximately 350 µg/l if derived from 0.75 litres of drinking water, very close to the worst-case average concentrations of lead predicted by plumbosolvency testing in the UK without corrosion inhibition. This suggests the potential for higher pre-natal mortality rates within a population if the drinking water is sufficiently plumbosolvent. Epidemiological research is needed to confirm the position.

Key points:

1. *Lead is toxic to humans and has no known biological role.*
2. *Lead poisoning is exhibited by a wide range of clinical conditions.*
3. *Lead in drinking water correlates with lead in blood in numerous studies.*
4. *Numerous studies correlate clinical effects with different concentrations of blood lead.*
5. *Reductions in the IQ of infants are of greatest concern.*

5 Evidence of problems with lead in drinking water

5.1 Emerging evidence of problems in Europe

Arising from the preliminary assessment of sampling methods (COST Action 637, 2008) the data for lead that has so far emerged, based on at least 100 samples in each case, is summarised in Table 5.1. The term "emerging evidence" is used as either the data has only recently been obtained or the data has not been previously published in a collated EU-wide format. In Table 5.1, with the exception of the data from Wales, all data presented was obtained between 2002 and 2008. The data from Wales derived from 1990 to 1995, prior to the commencement of ortho-phosphate dosing to reduce plumbosolvency; it is typical of the UK more generally (prior to ortho-phosphate dosing) based on other case studies (for example: Hayes *et al.*, 2006) and has been included for comparison.

This emerging evidence of non-compliance is not generally evident in the recent synthesis report from the European Commission (2008) for the period 1999 to 2001, albeit four countries were identified where lead problems existed; however, some countries did not submit their national report, some national reports were incomplete, the methods of sampling used for lead were not identified, and the bench-mark for compliance assessment was the earlier standard for lead of 50 μg/l from Directive 80/778/EEC.

It can be concluded that non-compliance with the lead standard of 10 μg/l (that will become a legal requirement in 2013) is significant in some parts of the EU and that plumbosolvency problems are not just confined to the UK (where comprehensive corrective action by water treatment has now been undertaken). Non-compliance with the interim standard of 25 μg/l is less pronounced but still conspicuous in some locations.

There is further evidence of problems with lead in Europe. Extensive laboratory based plumbosolvency testing (see Section 10.2) in the UK using the method of Colling *et al.* (1987) indicates (Hayes, 2008) that most, and possibly all, types of drinking waters in supply are likely to be sufficiently plumbosolvent so as to cause non-compliance with both the EU standards for lead (i.e.: 10 and 25 μg/l) wherever lead pipes are present (if corrosion inhibitors are not dosed). This data is summarised in Table 5.2 for the treated water from 158 water treatment works in the UK (obtained over the period 1999 to 2004), prior to any dosing of corrosion inhibitor, at the test temperature of 25°C, for three simplified categories.

The median 30 minutes contact (30MC) data for each sample tested, when adjusted for temperature, equates to a 30 minutes stagnation (30MS) sample for water in contact with a lead pipe in a water supply (assuming no dilution from water stood in non-lead pipe-work). If the 30MC test data at 25°C is halved, it equates (Croll, 2000) to the annual average temperature of water supplies in the UK (12 to 13°C). This temperature adjustment will be less for countries where the annual average water supply temperature is higher.

Work on water use patterns (Lacey and Jolley, 1986) has suggested that 30 minutes is a reasonable estimate of the average water-pipe contact time in a domestic dwelling, inferring that 30MS is a reasonable estimate of the average concentration of lead emitted from a tap (assuming no dilution from water stood in non-lead pipe-work). On this basis, not only were all 158 test waters found to be capable of exceeding the EU's lead standard of 10 μg/l, that is based on a weekly average concentration, most were found to also be capable of exceeding the interim EU standard of 25 μg/l.

Table 5.1. Emerging data on lead in drinking water (from Hayes and Skubala, 2009b)

Country	Area	Sampling method[a]	% samples >10 μg/l Pb	% samples >25 μg/l Pb	Source
Austria	All	FD	–	7.5	Jung and Heiss (2007) and COST Action 637 (2008)
	Vienna	FD	–	18.6	
Czech Republic	All	RDT	0.4	0.2	Němcová (2008)
Germany	N. Rhine Westfalia[b]	RDT	6.1–10.3	3.0–3.5	Ruebel and Becker (2008) and COST Action 637 (2008)
	Lower Saxony[b]	RDT	3.9–10.3	1.5–2.9	
	Bavaria	RDT	4.3	0.0	
Italy	All except Sardinia	RDT[c]	2.0	–	Veschetti (2008) and COST Action 637 (2008)
Netherlands	All	RDT	2.0	0.6	Slaats (2008) and COST Action 637 (2008)
	The Hague	RDT	23.0	–	Van Dongen *et al.* (2008)
Poland	All	FF	3.28	0.67	Witczak and Postawa (2008) and COST Action 637 (2008) (updated with 2007 data)
Wales (UK)[d]	All – pre PO_4 dosing	RDT	18.7	–	Hayes *et al.* (2008)

[a] FD = first draw; RDT = random daytime; FF = fully flushed.
[b] Results from different surveys.
[c] Likely to be an under-estimate, because of the way sampling locations were selected.
[d] Following the commencement of ortho-phosphate dosing in Wales, less than 1% of RDT samples now exceed 10 μg/l (Hayes *et al.*, 2008).

Table 5.2 Summary of plumbosolvency testing results (from Hayes, 2008)[a]

Type	Alkalinity	N	Ave pH	pH range	Ave median 30MC Pb (μg/l)	Range in median 30MC Pb (μg/l)	Ave 16hr stagnation Pb (μg/l)	Range in 16hr stagnation Pb (μg/l)
Ground	High	47	7.8	7.2–8.3	66	23–167	254	65–860
Surface	High	10	7.7	7.3–8.3	97	62–151	442	222–750
Surface	Low	101	7.8	6.7–8.8	172	42–694	855	109–3350

[a] 30MC refers to 30 minutes contact between water and lead piping during testing.

The lead concentrations after 16 hours stagnation give an indication of the equilibrium concentrations of the test waters when in contact with lead piping. Halving the test data (temperature adjustment) indicates that concentrations as high as 1675 μg/l could be expected from consumers' taps in the UK after overnight standing (assuming no dilution from water stood in non-lead pipe-work and prior to corrective action).

Further evidence of the potential scale of problems with lead in drinking water can be inferred from the pan-European study on lead monitoring reported by van den Hoven *et al.* (1999) that involved France, Germany, the Netherlands, Portugal and the UK. Split-flow composite (COMP) sampling was undertaken at 289 houses across 11 water supply zones, of which two were phosphate dosed and one partly phosphate dosed. Sixty-nine percent of the houses had a lead service pipe or internal lead piping. COMP provided a direct measure of the weekly average lead concentration at the houses investigated. The 90th percentile concentration determined in each zone varied from 9.6 to 107.8 μg/l, with 44% of the houses investigated exceeding the EU's standard of 10 μg/l.

It is interesting to note here the relationship between the percentage of houses that exceeded the EU's lead standard of 10 μg/l and the percentage of houses supplied by a lead pipe, In the van den Hoven *et al.* (1999) study, 64% of the houses with a lead pipe exceeded the standard, based on COMP sampling. This compares very closely to the 67% of houses with a lead pipe that were found by RDT sampling to exceed 10 μg/l in a major plumbosolvency control programme (Hayes *et al.*, 2008) in Wales (UK) prior to phosphate dosing, further supporting the view (van den Hoven *et al.,* 1999) that RDT and COMP are equivalent in zonal assessment terms (if the zones are mostly without corrosion inhibitor).

Taken together, these two studies suggest that around 65% of houses that have a lead pipe, either as a connection pipe or as internal piping, can be expected to exceed the EU's lead standard of 10 μg/l, unless corrosion inhibitor is dosed (and correctly so), which in the EU is not common outside the UK. This raises the question: how many houses are supplied by a lead pipe?

5.2 Occurrence of lead pipes in the EU

The data on the occurrence of lead pipes in the EU is poor, with the most recent assessment by van den Hoven *et al.* (1999) being summarised in Table 5.3. The basis of this data is not known, but it is likely to be no more than a collation of best estimates. These estimates suggest that about 25% of houses in the EU may have a lead pipe, potentially putting 120 million people at risk in today's 27 Member States. There is obvious scope to improve knowledge of the occurrence of lead pipes in the EU.

Table 5.3 Occurrence of lead pipes in Europe (from van den Hoven *et al.*, 1999)

Country	% Pb communication pipes	% Pb supply pipes or internal Pb plumbing
Belgium	19	15 to 30
Denmark	0	0
France	39	38
Germany	3	9
Greece	<1	0
Ireland	50	51
Italy	2 (?)	5 to 10
Luxembourg	7	0
Netherlands	6	8
Portugal	?	32
Spain	>3 (?)	?
UK	40	41

RDT sampling has been used in the UK to assess compliance with lead at consumers' taps since 1989 and in recent years many UK water companies increased their monitoring frequencies for lead as part of a major plumbosolvency control campaign prompted by the UK Government (for England and Wales Drinking Water Inspectorate, 2000, 2001). As bacteriological sampling frequencies are approximately 10 to 20 times higher than for trace metal "audit" samples, and as most of the costs of monitoring derive from sampling logistics, the opportunity was taken to take samples for lead analysis at the same locations being sampled for bacteriological parameters.

Assuming that no significant operational changes have taken place, it is entirely reasonable to bulk data from several years to increase the size of the data-sets. By determining the percentage of samples in which lead was detected analytically (typical limit of detection 1 μg/l or less) it has been possible to gain an estimate of the percentage of houses with a lead pipe.

This simple approach was used in the plumbosolvency control programme in Wales (Hayes *et al.*, 2008) and enabled a zonal emission model to be calibrated; that good validation of predicted RDT sample results was obtained from actual RDT sample results indicated that this simple methodology was adequate for estimating the percentage of houses with a lead pipe. However, to put this into perspective, over 11,000 results were available across 29 water supply systems, with an average of 383 RDT sample results per scheme.

5.3 Occurrence of lead pipes in the North America

Based on limited data, at least 260,000 lead service connections are believed to exist in Canada. Estimated in 1990, there were believed to be about 3.3 million lead service connections in the US (AWWA, 1990). For North America (excluding Mexico) these estimates equate to between 2 and 3 % of dwellings having a lead service pipe, significantly lower than in Europe.

Key points:

1. *Virtually all drinking water is sufficiently plumbosolvent to cause exceedence of modern day standards for lead in drinking water, if corrosion inhibitor is not dosed and/or if pH corrections are not applied.*
2. *Approximately 25% of houses in Europe are likely supplied by a lead pipe.*
3. *In North America, between 2 and 3% of houses are likely supplied by a lead pipe.*
4. *Two-thirds of the houses supplied by a lead pipe will likely fail the WHO Guideline Value of 10 μg/l for lead in drinking water, unless corrective treatment is applied.*

6 Risk assessment and health surveillance

6.1 Defining risks from lead in drinking water

There are a number of ways in which the risks from lead in drinking water can be defined, either for an entire water supply system or for an individual house, on the basis of:

(1) Quantification of the severity and likelihood of problems occurring, using a risk assessment matrix, as part of a Drinking Water Safety Plan.
(2) Compliance with regulatory standards.
(3) Exposure to lead concentrations that are considered to be of health concern.

The World Health Organization (2004) has suggested, as an example, a simple risk scoring matrix, as shown in Figure 6.1:

Semi-quantitative scoring allows control measures to be ranked in relation to the most significant hazards.

The following working definitions are suggested, each with a numerical score:

Likelihood:	Definition:	Score:
Almost certain	Once per day	5
Likely	Once per week	4
Moderately likely	Once per month	3
Unlikely	Once per year	2
Rare	Once every five years	1

Severity:	Definition:	Score:
Catastrophic	Potentially lethal to large population	5
Major	Potentially lethal to small population	4
Moderate	Potentially harmful to large population	3
Minor	Potentially harmful to small population	2
Insignificant	No impact or not detectable	1

The World Health Organization suggests that a matrix score of 6 should be used as the trigger for improvement action. This score results from two combinations:

(1) "Moderate severity (3)" and "Unlikely (2)", and
(2) "Minor severity (2)" and "Moderately likely (3)".

Likelihood	Insignificant	Minor	Moderate	Major	Catastrophic
Almost certain					
Likely					
Moderately likely					
Unlikely					
Rare					

Figure 6.1 Risk assessment matrix

In relation to lead in drinking water these combinations seem appropriate but there is, as yet, no consensus on how to define "potentially harmful" nor how to convert likelihood, when expressed as a frequency of occurrence, to zonal compliance with regulatory standards. Consensus will only come from case studies and peer review.

In the mean time, "potentially harmful" is defined in this Guide as either:

(1) Non-compliance with the WHO Guideline Value of 10 μg/l for lead in drinking water, or
(2) Exceedance of the US trigger for health concern of 10 μg/dl for lead in blood.

There are presently three reference points for converting zonal compliance into the term "likelihood" or related grades of concern:

(1) In the UK, one of the criteria for optimising ortho-phosphate dosing was that no more then 2% of RDT samples should exceed 10 μg/l; in practice, in England and Wales as a whole, a failure rate as low as 1% has been achieved.
(2) In the US, the 90th percentile concentration from survey samples must comply with 15 μg/l, on the basis of minimum 6 hour stagnation samples; this should equate to 90% of samples complying with this standard.
(3) In the guidance (Hoekstra *et al.*, 2009) provided to the Parties to the Protocol on Water and Health on sampling and monitoring lead in drinking water, it was proposed that the results from RDT sampling surveys should initiate the different grades of response shown in Table 6.1.

Table 6.1 Priorities for attention dependent on RDT survey results

Percentage of samples exceeding 10 μg/l	Priority for attention
<2.0	Low priority
2.0 to <5.0	Investigate any localised clusters
5.0 to <10	System-wide measures may be required in addition to resolving any localised clusters
10 to <20	System-wide measures required
20 to <50	Significant problems require attention
>50	Very significant problems require urgent attention

It is reasonable to propose that the zonal failure rate vs 10 μg/l, based on RDT sampling, that triggers corrective action is 2%, as this is also the UK optimisation target (Drinking Water Inspectorate, 2000, 2001). However, if it is assumed that an average water lead concentration of 20 μg/l is equivalent to a blood lead concentration of 10 μg/dl (Quinn and Sherlock, 1990), the comparable zonal failure rate, based on RDT sampling and 20 μg/l, that triggers corrective action would then be about 1%. In both cases, the failing RDT samples should be investigated to determine if localised action is required.

6.2 Assessing risk at the zonal level

As discussed in Chapter 3, it is not logistically possible to survey a water supply system using proportional sampling to determine average lead concentrations. It is, however, possible to predict such concentrations by computational modelling (see Section 10.3). Modelling can also predict the results of random daytime sampling if sufficient calibration data is available. This has potential benefit as RDT monitoring should normally take at least one full year to establish a zone's position, and data bulked from several years will reduce reproducibility problems.

Using the computational models, it is possible to investigate the extent of lead emissions across a City or Town for a wide range of plumbosolvency conditions, for different extents of occurrence of lead pipes and for different risk benchmarks. In a recent study (Hayes, 2010), the range of plumbosolvency conditions investigated were those shown in Table 6.2 for extents of occurrence of lead pipes ranging from 10 to 90%.

Table 6.2 Plumbosolvency categories (from Hayes, 2010)[a]

Category	M (μg/m²/s)	E (μg/l)	Typically associated water quality characteristics
Very high	0.3	450	Very low alkalinity, coloured >10°H (humic and fulvic acids), poor pH buffering capacity with tendency for pH to fall below 7.0. Particularly the case when water treatment is very limited.
High	0.2	300	Surface derived waters with a total organic content >3 mg/l, including high alkalinity waters affected by algae or containing treated sewage effluent. Particularly the case where water treatment is based on slow-sand filtration or where physico-chemical treatment is limited or poorly controlled.
Moderate	0.1	150	Surface derived waters after full physico-chemical treatment (including pH adjustment to 7.5–8.5) and most ground-waters.
Low	0.06	90	Some high quality chalk ground-waters
Ortho-phosphate dosed	0.02	30	Must be correctly dosed. Required doses tend to range from 0.5 to 2.0 mg/l (P) and must be applied consistently.

[a] M = initial mass transfer rate of lead into water. E = lead equilibrium concentration.

Two risk benchmarks that have been modelled are:

Benchmark 1: Percentage zonal exceedance of 10 μg/l, on the basis of simulated RDT sampling

The results from Hayes (2010) are shown in Table 6.3.

Table 6.3 Predicted risks using benchmark 1: 10 μg/l based on RDT sampling (from Hayes, 2010)[a]

Plumbosolvency			Percentage houses in zone >10 μg/l based on RDT samples, for each %Pb occurrence				
Category	M	E	10% Pb	30% Pb	50% Pb	70% Pb	90% Pb
Very high	**0.3**	**450**	6.5	18.9	31.6	45.1	56.6
High	**0.2**	**300**	5.2	16.7	28.0	38.7	49.0
Moderate	**0.1**	**150**	3.9	12.1	20.2	28.9	37.0
Low	**0.06**	**90**	2.5	7.7	13.5	18.4	23.5
Phosphate Dosed	**0.02**	**30**	0.4	1.1	2.1	2.7	3.2

[a] Zone modelled had 10,000 houses and was calibrated using the statistical distributions shown in Figure 10.7. RDT results were based on the average of 100 simulated surveys each of 100 samples. M and E define the plumbosolvency of the water (see Section 10.3 for a detailed explanation).

Table 6.3 has a number of important features:

(1) It indicates that in the worst case modelled (very high plumbosolvency and 90% of houses with a lead pipe) 56.6% of the houses in the zone were predicted to exceed the benchmark for lead in drinking water, equivalent to the percentage of the zone's population at risk on this basis.
(2) The UK's trigger (Drinking Water Inspectorate, 2001) for optimising corrective water treatment to reduce plumbosolvency was 5% or more RDT samples exceeding 10 μg/l. It can be seen that this level of non-compliance was exceeded for most of the cases modelled.
(3) The phosphate dosed condition that was modelled (M = 0.02, E = 30) was predicted to achieve or only slightly exceed one of the UK's optimisation criteria (Drinking Water Inspectorate, 2001) for ortho-phosphate dosing, that no more than 2% of RDT samples should exceed 10 μg/l. This dosed condition has readily been achieved in practice and, where necessary, lower values of M and E have been achieved by slightly higher phosphate doses in order to meet the 2% RDT target (Hayes *et al.*, 2006, 2008).

Benchmark 2: Percentage zonal exceedance of 20 μg/l, on the basis of simulated daily/weekly average lead concentrations (DAC) at all simulated houses

The results from Hayes (2010) are shown in Table 6.4.

The rational of the second benchmark is the approximate relationship between an average water lead concentration of 20 μg/l and a blood lead concentration of 10 μg/dl, the trigger for health protection in the US, based on the review of Quinn and Sherlock (1990).

In Table 6.4, the same trends are evident when compared to benchmark 1 except that the zonal failure rates are slightly lower, a reflection of the higher benchmark value. It can also be noted that the phosphate dosed conditions were found to have a 500-fold less failure rate than the very high plumbosolvency conditions and a 300-fold less failure rate than the high plumbosolvency conditions, for all percentages of houses with lead pipes. This clearly demonstrates the very substantial reduction in risk that can be achieved by ortho-phosphate dosing.

Table 6.4 Predicted risks using benchmark 2: 20 μg/l based on average lead concentrations (DAC) – from Hayes, 2010[a]

Plumbosolvency			Percentage houses in zone > 20 μg/l based on average lead concentrations, for each % Pb occurrence				
Category	M	E	10% Pb	30% Pb	50% Pb	70% Pb	90% Pb
Very high	**0.3**	**450**	5.4	15.8	26.5	37.0	46.9
High	**0.2**	**300**	3.1	9.5	16.0	22.7	29.4
Moderate	**0.1**	**150**	1.2	3.6	5.8	8.5	10.7
Low	**0.06**	**90**	0.4	1.2	2.1	2.6	3.2
Phosphate Dosed	**0.02**	**30**	0.01	0.01	0.02	0.05	0.11

[a] Zone modelled had 10,000 houses and was calibrated using the statistical distributions shown in Figure 10.7. Daily average concentrations were calculated for all simulated houses.

On the basis of extensive laboratory plumbosolvency testing, most treated drinking water without corrosion inhibitor has a moderate to high plumbosolvency, although there are many cases of waters having a high to very high plumbosolvency. The typical water qualities associated with plumbosolvency are listed in Table 6.2 and the occurrence of this range in plumbosolvency is shown in Table 6.5 for 45 zones from published case studies in the UK (Hayes, 2002; Hayes *et al.*, 2006, 2008), providing further evidence that the levels of zonal risk shown in Tables 6.3 and 6.4 are realistic.

Table 6.5 Plumbosolvency of drinking water in 45 UK zones prior to ortho-phosphate dosing

Plumbosolvency category based on factor M	Percentage of zones
Low to moderate: 0.050 to 0.099	35.6
Moderate: 0.100 to 0.149	42.2
Moderate to high: 0.150 to 0.199	11.1
High: 0.200 to 0.249	11.1

An assessment of the occurrence of lead piping in Europe (van den Hoven *et al.*, 1999) suggests that over-all about 25% of houses have a lead pipe (Hayes and Skubala, 2009a,b) but it is likely that the percentage of houses with a lead pipe is much higher in the older districts of many Towns and Cities. In consequence, the levels of zonal risk shown in Tables 6.3 and 6.4 are considered to be a fair reflection of the current circumstances in many European Towns and Cities where corrosion inhibitors are not dosed to the water supply, for the two risk benchmarks investigated. This is also borne out by the results of real-world RDT sampling, as illustrated in Table 6.6 for the 45 UK zones referred to above. The 45 zones summarised in Table 6.6 include both urban and rural areas and the failure profile is therefore probably slightly optimistic in relation to older urban areas alone.

The profile of observed zonal failure rates (Table 6.6) is also consistent with observations in France (Baron, 2001) in which zonal failure rates against 10 μg/l based on RDT sampling ranged from 11 to 57% across 7 study areas.

Table 6.6 Zonal failure rates for lead in drinking water in 45 UK zones, based on RDT sampling and exceedance of 10 μg/l, prior to orthophosphate dosing

Percentage RDT samples > 10 μg/l	Percentage of zones in each category
0 to 9.9	37.8
10 to 19.9	31.1
20 to 29.9	20.0
30 to 39.9	6.7
40 to 49.9	4.4

Rapid method for undertaking an initial risk assessment for a water supply zone

The predictions of zonal failure shown in Tables 6.3 and 6.4 can be used as a rapid initial basis for undertaking a risk assessment, as outlined below.

Step 1: Estimate the percentage of houses in the water supply zone that are likely to have lead piping, either to connect to the water main or within the house, or both. One approach is to consider the percentage of houses built at least 30 years ago (Hoekstra *et al.*, 2009), after which lead piping was generally no longer used in Europe.

Step 2: Select the plumbosolvency category that best fits the quality of the drinking water in the water supply zone from Table 6.2.

Step 3: Estimate the level of risk from Tables 6.3 and 6.4, using the percentages of houses that exceed the various benchmarks.

As sufficient RDT sampling results become available, Table 6.3 can be used, interpolating where necessary, to provide an indication of likely plumbosolvency conditions for any estimated percentage of houses with a lead pipe. The other risk assessment in Table 6.4 can easily be cross-referenced to that obtained from Table 6.3.

Method for undertaking a more accurate risk assessment for a water supply zone

A more precise estimate of risk can be obtained by the following steps.

Step 1: Determine the percentage of houses in the water supply zone that have lead piping, either to connect to the water main or within the house, or both, based on an inspection survey of a representative number of houses.

Step 2: Determine the plumbosolvency of the treated water input(s) to the water supply zone and a representative number of locations within the zone; this will typically require five samples to be tested using the established laboratory procedure of Colling *et al.* (1987). Test results can be obtained within one month.

Step 3: Use the zonal model to determine the level of risk. Calibration can be based on the data from Steps 1 and 2, and the assumptions illustrated in Section 10.3 (Figure 10.6). Alternatively, these assumptions can be fine-tuned in the light of local information, much of which can be gathered from housing surveys and water consumption data.

Step 4: It is preferable that RDT sampling has also been undertaken, to provide direct validation of the model's predictions.

The great benefit of fine tuning and validating the model in relation to local circumstances is that the model can then be used to assess the likely result of corrective actions in a local context.

6.3 Risk assessment at individual premises

Computer simulation, based on lead dissolution from lead pipes, can predict the average concentration of lead in drinking water, if sufficient calibration data are available, and is illustrated by Table 6.7.

Table 6.7 Predicted average lead (DAC) emissions from a single lead pipe (from Hayes, 2010)[a]

Plumbosolvency			Average Pb concentration (μg/l) from a range of lengths of lead pipe				
M	E	10 m	20 m	30 m	40 m	50 m	100 m
0.3	450	14.5	28.9	42.8	56.5	69.3	128.0
0.2	300	9.7	19.3	28.5	37.7	46.2	85.3
0.1	150	4.8	9.6	14.3	18.8	23.1	42.7
0.06	90	2.9	5.8	8.6	11.3	13.9	25.6
0.02	30	1.0	1.9	2.9	3.8	4.6	8.5

[a] Assumptions: daily water consumption = 450 L; water use pattern A (Figure 10.7) with half-hourly flow events; flow rate = 0.1 L/s; internal pipe diameter = 12 mm.

It can be seen from Table 6.7 that an average lead concentration of 10 μg/l or greater is predicted to occur with fairly short lengths of lead pipe for most plumbosolvency conditions. The critical lead pipe lengths increase with decreasing plumbosolvency. Phosphated water (M = 0.02; E = 30) does not approach the risk level with lead pipes up to 100 m in length, for lead pipes of 12 mm internal diameter. Except with very highly plumbosolvent water, lead pipe lengths less than 10 m (12 mm internal diameter) are not expected to pose a risk, as defined by exceedance of 10 μg/l based on average lead concentrations.

Table 6.7 can be used as a simple look-up table if estimates of lead pipe length and plumbosolvency category (Table 6.2) are known. However, it should be appreciated that:

- larger lead pipe diameters increase average lead concentrations for a given water volume and a given lead pipe length, due to a greater surface area of lead piping;
- average lead concentrations increase with lower water volumes;
- other water use patterns will have slightly different lead emission characteristics.

For these reasons, a more accurate risk assessment will be obtained if pipe-work characteristics and water consumption have been determined by inspection and if the plumbosolvency of the water in the supply zone has been determined by testing.

6.4 Health impact assessment

The benchmarks used in Tables 6.3 and 6.4 to assess zonal risk have one major limitation: they assume risk in zonal terms if the benchmark value is exceeded and do not consider the severity of the risk. This pass/fail basis is the common approach taken with many drinking water quality standards. In practice, some consumers will be at greater risk than others, depending on the concentrations of lead that they are exposed to from their drinking water. For example, the risk to health from an average lead concentration of 50 μg/l is clearly likely to be greater than from an average lead concentration of 11 μg/l. However, the risk to consumers from 9 μg/l is unlikely to be much different to that from 11 μg/l (for any given consumption and human tolerance), albeit these two concentrations are significantly different in terms of compliance/failure with the WHO Guideline Value.

The computational modelling methods described can investigate zonal exposures for any given range of benchmark values and it is possible to produce a zonal risk profile, as illustrated in Table 6.8 for the high plumbosolvency condition and 70% houses with a lead pipe. Such profiles will enable a better appreciation of the severity of the risks within a zone.

Table 6.8 Illustration of a zonal risk profile (from Hayes, 2010)[a]

Benchmark value, based on average lead concentrations – DAC (μg/l)	Predicted zonal failure/population exposed (%)
10	43.7
20	22.6
30	13.2
40	8.5
50	5.2
75	2.4
100	0.9

[a] Based on highly plumbosolvent water (M = 0.2, E = 300) and 70% of houses with a lead pipe. All other assumptions as in Figure 10.7.

The risk assessment methods that have been illustrated will enable the possible scale of problems with lead in dinking water to be quantified for the purpose of informing policy and for guiding improvement priorities. The methods are complimentary to actual zonal sampling and enable risk assessments to be undertaken very quickly.

Whilst zonal risk assessment will guide improvement planning, the direct determination of specific health risks to individual consumers will, at the minimum, require blood lead surveillance (as already practiced in the US). Epidemiological assessment of the populations deemed to be at risk should also be considered.

Key points:

1. *At the zonal level, risk from lead in drinking water can be defined by compliance with the WHO guideline value of 10* μg/l on the basis of random daytime sampling.
2. *Alternatively, a bench mark of 20* μg/l can be used, on the assumption that it approximates to the US blood lead trigger for action of 10 μg/dl.
3. *Computational modelling predicts that zones with a moderate plumbosolvency and more than 30% of houses with a lead pipe will require zone-wide action.*
4. *Zonal failure profiles give a better insight into likely health significance.*
5. *The individual houses at risk will mostly have a lead pipe at least 10 metres long if the internal pipe diameter is 12 mm.*
6. *Optimal ortho-phosphate dosing will reduce risk by up to 500 times.*

7 Lead pipe replacement and other engineering options

7.1 Survey techniques for identifying lead pipes

House age and lead pipe usage period

The occurrence of lead pipes is usually associated with the age of the buildings. Generally buildings that are at least 30 years old can be suspected to have lead pipes. However, the period in which lead pipes have been used may vary between countries. Whilst lead pipes are the most significant source of lead in drinking water, an inventory can be extended to lead-containing solders, brass fittings, galvanised steel pipes/fittings and unplasticised PVC containing lead-stabilisers.

Plumbosolvency maps

The information of the age of the housing stock and the period that lead pipes have been used, can be combined on a plumbosolvency map of each water supply zone. The most basic map will indicate areas where lead pipes are expected and where not. The more details that are available on other lead-releasing materials, the more detailed will be the maps that can be produced.

The plumbosolvency maps can be refined by including information on the composition of the water that is supplied. Plumbosolvency is promoted by low alkalinity (<50 mg/l $CaCO_3$), low pH (particularly below 8), high organic content (colour $>10°$ Hz) and discoloured (particularly reddish iron-containing) water.

More detailed housing inspection

Pipe-work surveys will be best undertaken by Water Inspectors with local knowledge of their water supply system, gained from repair and maintenance work.

For a representative number of properties, details of pipe materials, diameters and lengths should be obtained, with excavation if necessary. The residency of each property inspected should be recorded and any available records obtained of metered water consumption.

Sequential sampling of residential properties

Successive draws of known volumes of water (preferably 250 to 1000 mL) can usually be effective in determining if lead pipes or lead-containing plumbing components exist at a given property. If the only objective is determining the presence of lead piping, a smaller number of larger volume samples is normally adequate. More numerous smaller-volume samples can be used for more precisely delineating lead contributing devices and materials in the premise networks (Britton and Richards, 1981; AwwaRF, 2008; Schock and Lytle, 2010).

 Best Practice Guide on the Control of Lead in Drinking Water. Edited by Dr Colin Hayes. ISBN: 9781843393696.
Published by IWA Publishing, London, UK.

7.2 Engineering options for lead pipe replacement or modification

There are three basic issues in lead pipe rehabilitation: (i) lead pipe lateral rehabilitation on the public portion of the water supply system that runs from the water main to the water meter or stop-cock (curb-stop); (ii) lead pipes on the private portion of the water supply line from the meter or stop-cock to the house; and (iii) lead pipe-work within the house. Whilst the only 100% control method will be the removal or all lead pipes, there are substantially less expensive alternatives available consisting of chemical treatment and internal cathodic protection. Total lead pipe removal also has the problem of split ownership and responsibility, disturbance, and destruction of land-scaping (external) and decoration (internal).

There are six basic technologies in use today for lead service pipe rehabilitation, namely: (i) open-trench, replacement along an existing route, (ii) replacement along a new route, (iii) pipe pulling of a new pipe, (iv) slip lining, (v) pipe coating, and (vi) internal cathodic protection.

Each of these technologies is discussed below along with their strengths and weaknesses:

Open-trench replacement along an existing route

This is a rapid method of repair that removes the existing lead pipe from service and replaces it with a non-lead pipe lateral such as copper or plastic.

- Advantages: complete removal of lead from the water supply piping system.
- Disadvantages: costly repair from a trenching standpoint as well as from the destruction of the paved surface in existence prior to the installation of the trench; must shut water off to make repair; may also be necessary to excavate a road or other highway, if the property is the opposite side to where the water main is located.

Directional drilling or "mole" replacement along a new route

This is a rapid method of repair that removes the existing lead pipe from service and replaces it with a non-lead pipe lateral such as copper or plastic.

- Advantages: complete removal of lead from the water supply piping system; less costly repair from a trenching standpoint due to need for smaller trench as well as from the lessened destruction of the paved surface in existence prior to the installation of the new pipe.
- Disadvantages – high costs and must shut water off to make repairs.

Pipe pulling replacement through an existing route

This is a rapid method of repair that removes the existing lead pipe from service and replaces it with a non-lead pipe lateral such as copper or plastic.

- Advantages: complete removal of lead from the water supply piping system; less costly repair from a trenching standpoint due to need for smaller surface opening, as well as from the lessened destruction of the paved or road surfaces in existence prior to the installation of the new pipe.
- Disadvantages – high costs and must shut water off to make repairs.

Slip lining replacement through an existing pipe

This is a rapid method of repair that inserts an inert lining in the existing lead pipe. It is not suitable for internal lead pipe-work and its application will be limited by valves and stop cocks. Trials in the UK were only partially successful (WRc, 1999).

- Advantages: removal of lead exposure to the water supply piping system.
- Disadvantages: unknown degree of success from any gaps in liner – pipe; must be reasonably clean prior to slip lining and difficulties with reductions in pipe diameter.

Epoxy lining through existing pipe

This is a method of repair that creates a lining after abrasively cleaning an existing lead pipe.

- Advantages: removal of lead exposure to the water supply piping system.
- Disadvantages: unknown degree of success due to any gaps in liner – pipe must be clean and dry prior to epoxy lining and there are difficulties with reduction in pipe diameter.

Insertion of an anode

This is a propriety non-replacement approach for lead pipe solubility reduction. It consists of inserting a platinum coated anode through existing piping to implement a rapid method of control of lead solubility by an impressed DC potential. Both sides of the meter assembly (when it is sandwiched between two lead laterals) can be protected.

- Advantages: internal cathodic protection provides a less intrusive approach than the complete removal of a lead pipe; it is a less costly repair as no significant trenching is required as well as less destruction of paved and road surfaces.
- Disadvantages: water may not need to be shutoff to install; however, performance is substantially improved with acid & peroxide cleaning to remove lead scale to bright lead surface – minor operating costs due to power for impressed current.

7.3 Partial lead pipe replacement

In late 2008, the AWWARF published "Contribution of Service Line and Plumbing Fixtures to Lead and Copper Rule Compliance Issues." This report found that from the sites in the US with lead services, that the lead service lines contributed from 50–75% of the lead in drinking water, onsite piping contributed from 20–35% of the lead in drinking water and that faucets contributed from 1–3% of the lead in drinking water. All other contributions, for the limited number of homes sampled, such as, the brass water meter fittings and the meter itself had lead contributions that were minimal.

In one City in Eastern England (UK), pipe inspections found that 52% of houses had a lead service pipe but only 5% had internal lead pipe-work. Elsewhere in Europe, it appears from Table 5.3 that the percentage of houses with internal lead pipe-work is much higher in some countries.

Partial lead pipe replacement of the public portion of the lead service lateral is being practiced in some cities. As found in recent US studies (AWWARF, 2008), partial lead pipe replacement is only a marginal strategy for controlling lead concentrations in drinking water at tap outlets. It is generally justified by reducing the liability of the water supplier for lead emissions from domestic pipes.

There are concerns that partial lead pipe replacement may actually worsen the situation in terms of lead emissions at the tap, based on the recent AWWA (2008) study. The removal of an existing lead lateral can disturb the complex lead hydroxyl-carbonate scale that has formed on the inside of lead pipes over years of service. Flaking of disturbed deposits can elevate particulate lead concentrations. Mechanical disturbance due to partial lead pipe replacement has been shown to elevate lead concentrations at the tap in several other studies (Britton and Richards, 1981; Breach *et al.*, 1991; Hulsmann, 1991; AWWARF, 1990; Wycock *et al.*, 1991; Boyd *et al.*, 2004) with high lead levels persisting for up to several months. In such circumstances, partial lead pipe replacement by water utilities may actual increase liability, particularly if adverse effects on consumers were demonstrated, for example by observations of increased levels of blood lead. Recently, an information notice from the US CDC has highlighted an epidemiological study in which elevated blood lead concentrations have been correlated with partial lead pipe replacement (CDC, 2010).

Depending upon the selection of materials, the introduction of a new copper pipe lateral provides a fresh surface for galvanic corrosion (corrosion where the lead pipe dissolves to protect the new copper pipe – as in a battery cell) and lead levels in the drinking water may increase substantially in the short term. The installation of a dielectric in this situation is recommended.

7.4 Total lead pipe replacement

It must be the ultimate goal to remove all lead pipes from service. This has been the strategy in the Netherlands, where substantial grant-aid was made available by the Government to encourage home-owners to replace their lead pipe-work coincident with the water company replacing its lead service line. Although in general this approach has proven successful, The Hague case study (Van Dongen *et al.*, 2008) demonstrates that the lack of interest and cooperation from home-owners may result in poor overall results. The principal consumer or home-owner concerns were the disruption and part costs that they would incur.

It is likely that legislation will be required to force home-owners to remove their lead pipes, perhaps at the time of sale or letting. This would likely take 25 to 50 years to become fully effective. In the US, a small number of cities have obtained sufficient legal power to require the replacement of the full service lines, including the part owned by the property owner. Examples are Madison, WI, and Lansing, MI, although their legal tools are different.

Key points:

1. *The simplest way to estimate the occurrence of lead pipes is housing age, from which plumbosolvency maps can then be drawn showing those areas at greatest risk.*
2. *More detailed inspection surveys are recommended.*
3. *There are six basic technologies in use today for lead service pipe rehabilitation, namely:* (*i*) *open-trench, replacement along an existing route,* (*ii*) *replacement along a new route,* (*iii*) *pipe pulling of a new pipe,* (*iv*) *slip lining,* (*v*) *pipe coating, and* (*vi*) *internal cathodic protection.*
4. *Partial lead pipe replacement will not solve the problem in many cases and can even make matters worse, at least in the short term.*
5. *Attempts at total lead pipe replacement that include internal domestic lead pipes have failed because of a lack of cooperation from home-owners.*

8 Corrective water treatment

8.1 pH elevation and centralised softening

Experience in the UK

The solubility of lead is reduced at high pH values, particularly for low alkalinity waters, as illustrated by Figure 1.2. This was exploited in areas such as Glasgow, Edinburgh and many parts of Wales (all UK) to reduce plumbosolvency. In many of these areas, compliance was mostly achieved with the earlier EU standard for lead in drinking water of 50 μg/l (which in the UK, was assessed on the basis of random daytime sampling from consumers' taps) by pH elevation to around 8.5 to 9.0, normally with lime. In Glasgow, supplementary ortho-phosphate dosing was required to minimise non-compliance with this earlier standard, as illustrated by Table 4.2; until the early 2000s the water supply to Glasgow did not receive comprehensive physico-chemical treatment and high organic colour was responsible for exacerbating plumbosolvency, explaining why ortho-phosphate dosing was initiated in the 1980s.

In the late 1970s, trials in Eastern England indicated that pH elevation of high alkalinity waters could at best achieve a 30% reduction in lead solubility. There were concerns about calcite precipitation and the approach was abandoned in favour of ortho-phosphate dosing. However, few high alkalinity supplies were ortho-phosphate dosed until the mid-1990s, when compliance with the former 50 μg/l standard was enforced.

With the advent of the current (25 μg/l from 2003) and future (10 μg/l from 2013) lead standards in the EU, 95% of the UK's public water supplies are now ortho-phosphate dosed, clearly accepting the limitation of pH elevation alone to achieve these modern day standards.

Lead solubility modelling indicates that the minimum solubility of lead (II) phosphate is approximately 7.5, implying that this pH value is the optimum condition for ortho-phosphate dosing. However, UK experience has been that (i) for high alkalinity waters and well treated low alkalinity waters, pH has little impact on the effectiveness of ortho-phosphate dosing within normal operating conditions (7.0 to 8.5) and (ii) for less well treated low alkalinity waters, it is better to maintain pH in the range 8.0 to 9.0, explained by organic effects.

Experience in the Netherlands

Since the 1980s, it has been common practice to increase pH to reduce plumbosolvency, allied to centralised water softening. Softening is commonly achieved using pellet reactor softeners with caustic soda. The approach has been generally successful in meeting the former lead standard of 50 μg/l but as the Hague case study demonstrates (van Dongen *et al.*, 2008) it can be insufficient for achieving the future lead standard of 10 μg/l.

Experience in North America

The vast majority of U.S. utilities were able to comply with the Lead and Copper Rule by: 1) pH and alkalinity adjustment, most frequently to the pH range of 9–9.5 for systems with extensive lead piping; 2) dosing of orthophosphate in the pH range of approximately 7.2 to 8; or 3) the formation of insoluble

PbO_2 deposits through chlorination to high ORP and sometimes concurrent with pH adjustment to >9. Once a utility has been brought into compliance, changes in water treatment that alter the corrosion chemistry, such as switching from free chlorine to monochloramine for secondary disinfection (Edwards, *et al.*, 2004) or increasing the chloride to sulphate mass (Triantafyllidou *et al.*, 2009) occasionally trigger massive lead release events. There are a surprising number of treatment changes that can influence corrosion chemistry in unexpected ways. Therefore, planning, including desktop and pilot studies, monitoring and vigilance should occur prior to making changes to the treatment.

For water systems with extensive lead piping, sufficient pH adjustment to avoid the use of phosphate inhibitors has essentially only been possible in rather low carbonate waters, where it is possible to achieve pH's in the range of 9.2–9.5, and sometimes higher. Complications to plumbosolvency control, both by pH adjustment and phosphate inhibitors, have also been found in water systems containing excessive aluminum, manganese, and/or iron, where phosphate dosing causes turbidity and clogging of various in-line flow control devices. The non-Pb coating material has been noted to consume orthophosphate and render it less effective at typical dosages. Phosphate dosing at typical US levels (0.3 to 1.0 mg/L as P) also has difficulty passivating the lead pipe and preventing sloughing off of lead-containing particulate material when these non-Pb surface deposits have been found to exist. Erratic Pb levels from particulate release are periodic problems in these systems.

As is predicted by solubility modelling (Schock *et al.*, 1996; AwwaRF, 1990; Schock and Lytle, 2010), the most effective pH range for orthophosphate dosing has been observed to be from approximately 7.2 to 7.8, depending slightly on DIC concentration. Numerous problems with controlling lead release at pH over 8 have been encountered, which is consistent with controlled laboratory experiments (Schock *et al.*, 1996; 2008)

8.2 The corrosion inhibitors used in reducing plumbosolvency

A number of chemical inhibitors are used in water supplies to reduce the corrosion of metals in contact with the water, mainly in the UK and US. These are considered below:

Silicate

Often added as sodium silicate, this chelating agent is able to keep ferrous ions in solution and avoid their oxidation to the insoluble ferric form of iron. Silicate can therefore reduce the extent of red water iron discolouration problems. Silicate does not reduce the dissolution of lead, but it may reduce particulate lead problems arising from iron discolouration.

Poly-phosphates

Poly-phosphates are used for corrosion control in some water distribution systems to reduce iron "red-water" discolouration, by the chelation of ferrous iron at the sites of active corrosion. Poly-phosphates also chelate other metals, including lead, and their dosing can be antagonistic to plumbosolvency control, as illustrated by Table 8.1. It can be seen that without ortho-phosphate, there is a significant increase in lead dissolution from the higher poly-phosphate dose.

Zinc phosphate

This inhibitor is not used in the UK for corrosion control as work by WRc in the 1980s showed that it was less effective than ortho-phosphate for reducing lead solubility. It is used in some US supplies to reduce cuprosolvency and for lead corrosion control. The zinc has some noteworthy benefits in preventing scaling and controlling concrete corrosion. The published US research has generally not found it more effective in the case of lead piping. There is still debate and research to be done with respect to whether the cathodic

Table 8.1 Effect of polyphosphate on the reduction in plumbosolvency by ortho-phosphate (from testing for a UK water service company)

	Median lead concentrations (μg/l) after 30 minutes contact with new lead pipe at 25°C		
Ortho-phosphate dose mg/l as P	Zero poly-phosphate	Poly-phosphate 0.2 mg/l as P	Poly-phosphate 1.6 mg/l as P
0	142	143	281
1	3	19	54
2	3	12	51
3	3	10	44
4	3	9	32

inhibition properties of the zinc are or may be useful in improving lead release from devices made from leaded brasses, or with lead-tin soldered joints.

Ortho-phosphate

Ortho-phosphate is dosed to 95% of public water supplies in the UK, most commonly in the acid form (normally food grade 75% w/w H_3PO_4) or in a few cases as a liquid form of the salt sodium dihydrogen phosphate, if less pH depression is required. It is the most effective inhibitor of plumbosolvency.

The success of this treatment is demonstrated by the 2007 and 2008 summary statistics for England and Wales (Table 8.2), based on random daytime sampling.

Table 8.2 Compliance with lead, copper and nickel standards in England and Wales in 2007 and 2008

Parameter	Standard	No. samples		%Compliance	
		2007	2008	2007	2008
Lead	10 μg/l	23,181	13,153	98.30	98.91
	25 μg/l	23,181	13,153	99.76	99.77
Copper	2 mg/l	4,661	5,659	99.89	99.95
Nickel	20 μg/l	12,856	12,451	99.84	99.84

Compliance with the lead standard of 10 μg/l is already >99% in some parts of England and Wales. It is worth noting that copper has not been considered a problem in the UK for many years, except for a few localised cases involving either intermittently low pH of the water in supply or poor installation involving debris. The high level of compliance with the EU nickel standard contrasts with lower levels of compliance elsewhere in Europe where ortho-phosphate is not dosed.

In the US, doses of ortho-phosphate, in any form, in excess of a 1 mg/l (P) residual, are often necessary to control cuprosolvency in high dissolved inorganic carbon (DIC >40 mg C/L) waters, particularly in new plumbing.

Ortho-phosphate suppresses plumbosolvency by converting some of the lead carbonate in the corrosion film to lead phosphate, which is less soluble. Within the three-dimensional corrosion film there is an equilibrium between lead phosphate and lead carbonate as determined by the relative concentrations of ortho-phosphate, carbonate and bicarbonate ions in the water within the lead pipes.

8.3 Engineering design of ortho-phosphate dosing plants

It is common to dose either food grade ortho-phosphoric acid (H_3PO_4) or food grade monosodium phosphate (MSP) (Na H_2PO_4) in the concentrated form supplied.

For larger supplies in temporate climates the use of 75% w/w ortho-phosphoric acid avoids solidification under freezing conditions. Supplies should be available in either 50 litre plastic carboys, 1 m^3 plastic intermediate bulk containers (IBCs) or in greater bulk amounts from a road tanker.

For larger water supplies, a direct feed system can be used which doses the chemical by a metering pump directly to the point of application, without carrier water. Dedicated storage tanks are generally required for these direct feed applications.

Small to medium sized water supplies may use carrier water systems. These systems may accept deliveries of chemical from carboys, Intermediate Bulk Containers (IBCs) or bulk tanker. The carrier water systems may be fully containerised or skid mounted and located within existing buildings. This type of dosing rig will enable 'standard' manufactured rigs to be used which can be installed within existing buildings which have some free space or in small secure kiosks which can be located outside existing buildings or on the roadside. System design for both chemicals are broadly similar and a generalised schematic is given in Figure 8.1.

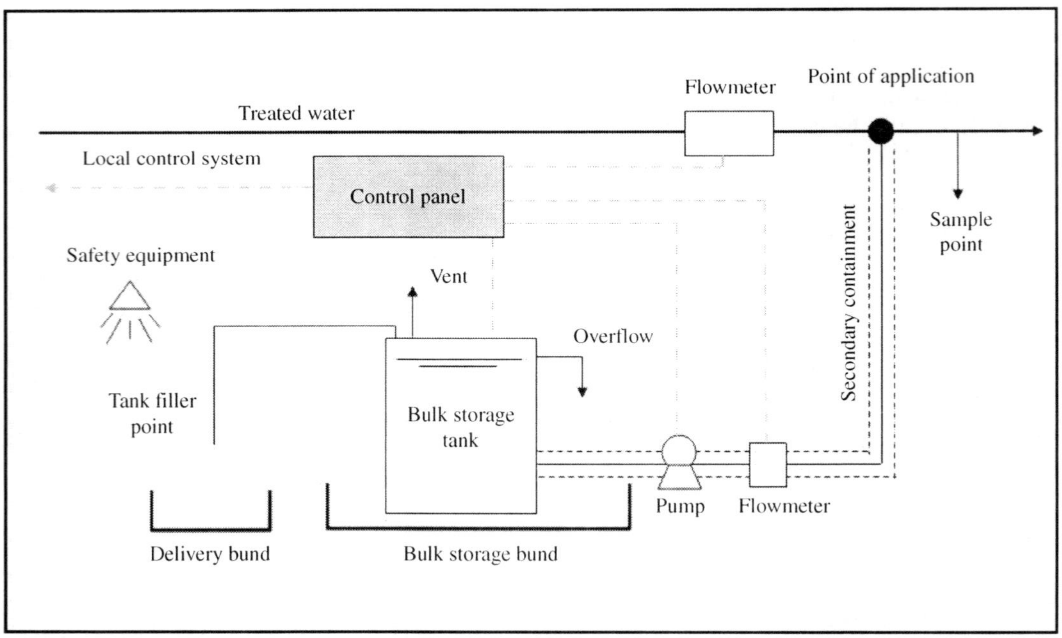

Figure 8.1 General diagram of an ortho-phosphate dosing plant

Storage of these chemicals can be inside or outdoors, the latter possibly requiring electric trace wire heating. All chemical storage must be bunded with secondary liquid containment equal to a minimum of 110% of the chemical volume held. Suitable health and safety provision, including a drench shower and eye wash station, must be available to plant personnel for any cases of chemical splashing. Spill containing equipment should also be available.

The general plant arrangement is for a duty dosing pump to inject the liquid chemical into a water transfer main, most commonly proportional to flow to ensure consistent dosing. A standby dosing pump should also be available, preferably with automatic switch-over if the duty pump fails.

The dosing plant will most commonly be located at a water treatment works but there are many examples of dosing plants being located within the distribution network in order to either dose just a local area or to dose water derived from multiple sources.

A positive displacement dosing pump should be used that is capable of delivering the amount of chemical required into the transfer main. It is essential that the injection lance has a non-return valve to prevent back-flow. The outlets from the lance should be positioned to minimise localised corrosion and to encourage rapid mixing, a static mixer often being used to assist with rapid mixing.

Other important design considerations include the provision of:

- A carrier water system (with break tanks) where necessary
- Non-return valves where appropriate
- Catch-pots to collect leaking liquids
- A calibration system for the dosing pump(s)
- Secondary containment of the pipe-work with leak detection
- Contained delivery areas and appropriate tanker access and drainage e.g. diversion to spill tank
- An accurate chemical flow-meter
- Level instruments for measuring liquid depths in storage tanks, bunds and catch-pots
- A control system with appropriate alarms
- Telemetry.

It is essential that dosing is proportional to water flow, particularly where flows vary. A signal from the flow meter will alter the pump stroke frequency on an automatic basis. Increasing or decreasing the dose applied can be achieved by manual setting of the length of the pump stroke. Ideally, the pump settings should be around 50% of the pumping range possible, and extreme pump settings avoided.

Some operators use an on-line ortho-phosphate monitor to record the phosphate concentrations achieved by dosing. It is important that sampling lines are located at a suitable down-stream position where mixing of the chemical is complete. The monitors typically sample every 15 minutes and use a colorimetric test procedure involving test reagents, which should be kept in date.

The signal from the phosphate monitor can be used to automatically trim dosing but experience in the UK has varied and many operators only use the monitors for local information, linking to alarm systems, validating performance and for water quality reporting.

Reliance is placed on accurate dosing of the chemical and therefore accurate flow measurement of the potable water and the chemical in direct feed systems. It should be noted that carrier water systems produce a system lag since the change in dose must be carried to the point of application within the carrier water before the dose into supply changes.

Where good flow metering is available for the main flow, it is feasible to consistently dose ortho-phosphate at a concentration within 10% of the target dose. Any greater deviation should be investigated.

The addition of ortho-phosphoric acid can cause a significant lowering of the pH of the potable water, particularly in soft waters with limited buffering capacity. The consequences of the changes in pH should be considered as part of the design process. The use of MSP may also cause a lowering of pH but the effects are not so pronounced. MSP is also easier to handle, with fewer health and safety issues.

8.4 Ortho-phosphate dose optimisation

The fundamental problems are that:

(1) Every drinking water supply system has its own individual characteristics (most importantly, the corrosivity of the water supply and the extent of occurrence of lead pipes), albeit generic groupings do help to develop corrective strategies and set priorities.

(2) There is no simple control loop available to use in the adjustment of ortho-phosphate doses, particularly when in-situ lead pipes can take two to three years to equilibrate with any given ortho-phosphate dose.

Optimisation based on random daytime sampling alone

A typical ortho-phosphate dose can be applied and its success (or otherwise) reviewed after a sufficient period of sampling. With groundwater supplies, a minimum period of sampling of one year may well be adequate. However, with surface derived waters, particularly those with a total organic carbon content greater than 3 mg/l, sampling for at least two years will likely be necessary.

After the sampling period, the ortho-phosphate dose can be adjusted (up or down) depending on the optimisation criteria that have been set. One of the UK's optimisation criteria was that no more than 2% of random daytime samples from consumers' taps should exceed a lead concentration of 10 μg/l.

The obvious problem is the time taken to achieve and demonstrate optimisation, bearing in mind the need for fairly large numbers of samples to characterise each dosing condition.

Pipe rigs and stagnation sampling

As described in Section 10.2, lead pipe test rigs can be installed at strategic locations within a water supply system. A typical deployment would be: (i) treated water at the treatment works, before ortho-phosphate dosing, (ii) treated water at the treatment works, after ortho-phosphate dosing, (iii) within the near distribution system, close to the treatment works, and (iv) within the far distribution system, distant from the treatment works.

If new lead pipes are installed in the rigs, they equilibrate quickly with their feed water. This indicates the ultimate performance potential of the applied ortho-phosphate dose but does not provide any indication of the time that in-situ lead pipes will take to reach equilibrium. If old exhumed lead pipes are installed in the rigs, there is a likelihood of erratic results caused by the physical disturbance of the lead corrosion deposits when exhumed and in subsequent handling.

A limitation of pipe rigs is that the results do not provide a measure of the extent of zonal lead emissions. Pipe rigs are more useful for demonstrating the on-going operational status of the ortho-phosphate dosing. An additional limitation of pipe rigs at the present time is that in the case of major phase transformations that might accompany the treatment change to be tested (for example, PbO_2 formation or breakdown), the effectiveness of new pipes has not been substantiated by research published so far, both in terms of sufficiently rapid formation of the relevant scale mineralogy, as well as the physical and chemical characteristics that can allow the impact of treatment changes to be adequately determined for the old pipes still in the ground. Additionally, the sensitivity to scale disruption and the optimization of handling techniques for exhumed service lines remains to be adequately investigated, in the absence of systematic validation of the use of new pipes and pipe rig operational protocols (scale aging, time, flow, stagnation events) for treatment change impact studies.

For simple Pb(II) chemistry systems as appear to dominate in the UK, where the only issue is determining the best ortho-phosphate dose for an unchanging water chemistry scenario, the use of new pipes is doubtless a cost-effective simulation. However, the US situation, with competing regulations requiring perhaps several significant water treatment process changes, plus possible Pb oxidation state changes, is more complicated at this time, and new pipe simulation may not be reliable without prolonged time periods that would be necessary to form analogous scales to those at the real residences and buildings.

Reference houses and stagnation sampling

Reference houses provide a direct measure of the response of in-situ lead pipes to ortho-phosphate dosing. They also provide evidence of the time being taken by in-situ lead pipes to achieve equilibrium (see Table 3.7). Monitoring at reference houses does not provide a measure of the extent of zonal lead emissions.

An integrated approach

The best and quickest way to achieve optimisation of ortho-phosphate dosing is to use several techniques in combination:

(1) Diagnostic analysis of available water quality data to determine possible influences of pH, organics and iron discolouration.
(2) Assessment of the likely extent of occurrence of lead pipes (Section 7.1).
(3) Laboratory plumbosolvency testing to directly determine the corrosivity of the water supply to lead pipes (Section 10.1).
(4) Inventory monitoring by random daytime sampling (Section 3.2).
(5) Zonal lead emission modelling to investigate corrective options and to determine ortho-phosphate dosing requirements (Section 10.3).
(6) Installation of lead pipe test rigs at appropriate locations (Section 10.2), or repeated stagnation sampling of suitable reference houses, or both.
(7) Operational monitoring by random daytime sampling to demonstrate that zonal objectives have been achieved.

The great advantage of this integrated approach is that a range of data types combine to provide confidence in the corrective measures undertaken, within the shortest possible timescale. The major case studies in the UK that used this integrated approach were very successful in achieving optimisation targets, within only two to three years.

Ortho-phosphate dose requirements

The ortho-phosphate doses needed to achieve the UK criterion, that no more than 2% of random daytime samples exceed 10 μg/l, varied from 0.5 to 2.0 mg/l (P) in the great majority of cases. Exceptionally, with poorly treated, highly coloured, low alkalinity waters, a dose of 2.8 mg/l (P) was found to be necessary. Table 8.3 illustrates the dose requirements for low and high alkalinity waters, based on extensive laboratory plumbosolvency testing. Such doses of ortho-phosphate are considered to be entirely safe, given that ortho-phosphate concentrations are many thousands of times higher in many carbonated soft drinks.

Table 8.3. Summary (based on a simple classification) of ortho-phosphate doses required to achieve a lead emission of <10 μg/l after 30 minutes contact at 25°C in laboratory tests (from Hayes, 2008)

Ortho-phosphate dose range (mg/l P)	Percentage of high alkalinity ground waters	Percentage of high alkalinity surface waters	Percentage of low alkalinity surface waters
0.5 to 0.6	6.5	0	33.7
0.9 to 1.2	60.9	26.7	38.6
1.4 to 2.0	32.6	73.3	23.8
2.2 to 2.8	0	0	3.9

In practice, the ortho-phosphate dose required is also a function of the percentage reduction in plumbosolvency needed to achieve the optimisation criterion, which is strongly influenced by the extent of occurrence of lead pipes within the water supply system. In the Wales (UK) case study (Hayes *et al.*, 2008), the required percentage reduction in plumbosolvency and the required average ortho-phosphate dose was established for a total of 29 dosing schemes that were subject to regulatory control, plus 9 others not subject to regulatory control.

The required percentage reduction in plumbosolvency averaged 76% but ranged from 50 to 92% as shown in Figure 8.2, the required reduction being strongly influenced by a combination of the plumbosolvency of the water prior to ortho-phosphate dosing and the percentage of houses in the supply area with a lead pipe. In this latter respect, the percentage of houses with a lead pipe was estimated to range from 8 to 65% with an average of 28%.

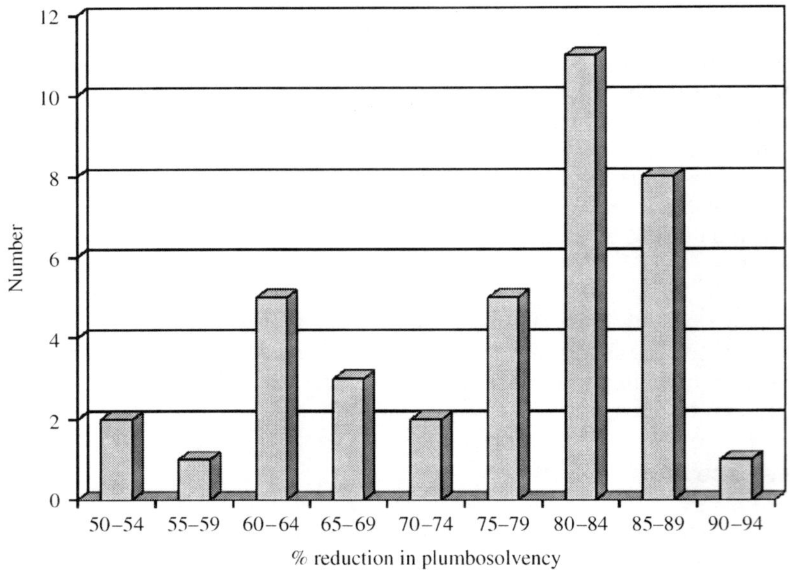

Note : The number of water supply systems requiring a particular reduction in plumbosolvency is shown for discretised ranges (for ease of presentation)

Figure 8.2 Required reductions in plumbosolvency in 38 water supply systems in Wales (UK) (from Hayes *et al.*, 2008)

The average dose of ortho-phosphate needed to achieve the <2% RDT sampling criterion in Wales ranged from 0.6 to 1.5 mg/l (P) with an over-all average of 0.9 mg/l (P).

The success of the integrated approach to plumbosolvency control in Wales is illustrated (Figure 8.3) by the random daytime sampling results before and after the introduction of ortho-phosphate to the 29 water supply systems subject to regulatory control.

Operational control

An on-line phosphate monitor can be used, with sampling after the phosphate injection point. Results can be displayed both locally and at a central control point (via telemetry). If preferred, the results can be used to control the dosing pump.

The phosphate set points of the dosing system can be set to achieve the average required dose but applied as a slightly higher summer dose and slightly lower winter dose. This arbitrary approach is considered to give better protection during the summer when lead concentrations are potentially at their highest. High-high, high, low and low-low alarm settings can be set to give two levels of alarm response; in general, high and low alarms prompt attention during the next site visit, whereas high-high or low-low alarms prompt immediate attention.

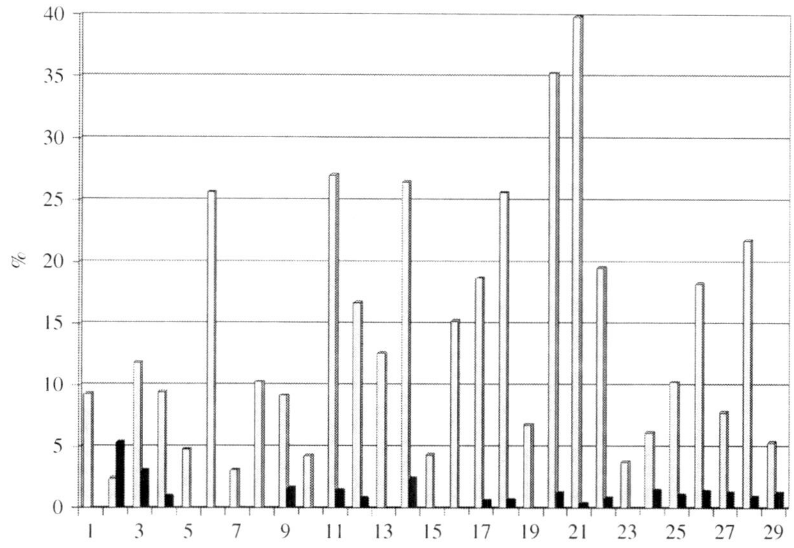

Key: light columns before dosing commenced, dark columns after

Figure 8.3 RDT sampling results before and after ortho-phosphate dosing in Wales (from Hayes *et al.*, 2008)

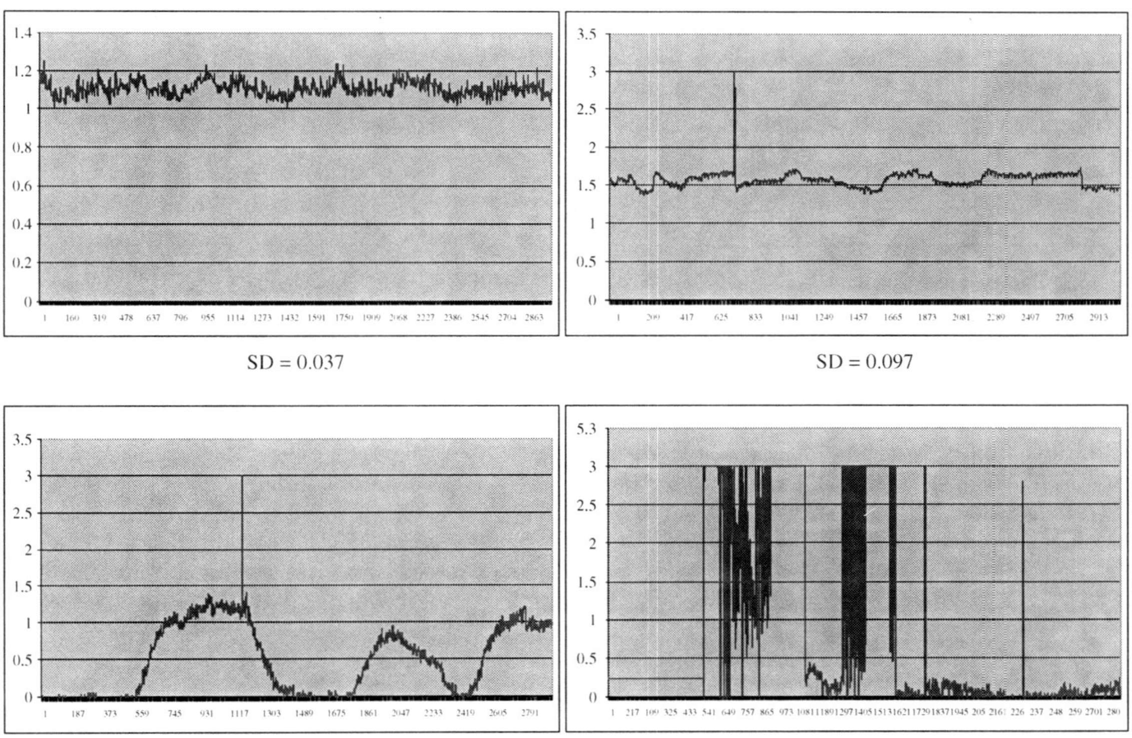

Figure 8.4 Examples of phosphate monitoring results (from Hayes *et al.*, 2008)

Table 8.4 Environmental impact assessment of dosing ortho-phosphate

Assumption	Calculations for impact condition 1 – no phosphate dosed to water supply	Calculations for impact condition 2 – phosphate dosed to water supply	Calculations for impact condition 3 – no phosphate dosed to water supply	Calculations for impact condition 4 – phosphate dosed to water supply
100,000 population and 150 litres per head per day	Base sewage flow = 15 Mld	Base sewage flow = 15 Mld	Base sewage flow = 15 Mld	Base sewage flow = 15 Mld
Phosphate concentration of sewage	10 mg/l (P)	11 mg/l (P) including 1 mg/l (P) dosed to water supply	10 mg/l (P)	11 mg/l (P) including 1 mg/l (P) dosed to water supply
Phosphate loading to sewage works	150 kg (P) per day	165 kg (P) per day	150 kg (P) per day	165 kg (P) per day
Phosphate elimination by sewage treatment	90%	90%	0%	0%
Phosphate loading from sewage works discharge	15 kg (P) per day	16.5 kg (P) per day	150 kg (P) per day	165 kg (P) per day
River provides 10 fold dilution of base sewage flow	River flow = 150 Mld	River flow = 150 Mld	River flow = 150 Mld	River flow = 150 Mld
Assumed phosphate concentration upstream of sewage works	0.10 mg/l (P)	0.10 mg/l (P)	0.10 mg/l (P)	0.10 mg/l (P)
Phosphate concentration downstream of sewage works	0.20 mg/l (P)	0.21 mg/l (P)	1.10 mg/l (P)	1.20 mg/l (P)

The phosphate monitors in use can experience problems, as a result of chemical reagent quality and monitor reliability. In the Wales case study, this prompted a comprehensive review leading to site-specific improvements where necessary and the adoption of a minimum acceptable criterion for monitoring that the standard deviation of results should not exceed 0.2 mg/l (P). The range in dosing performances that were experienced is illustrated by Figure 8.4, but were soon improved where necessary; it was found that the application of ortho-phosphate at a constant or near constant dose is necessary to maximise its effect.

8.5 Environmental impact of ortho-phosphate dosing

Ortho-phosphate dosing of water supplies is extensive in the UK and commonplace in the US, but elsewhere it is very limited. This is due to an under-estimation of the extent of problems with lead in drinking water and concerns about environmental impact. In Table 8.4, a simple environmental impact assessment is illustrated for situations where ortho-phosphate is dosed to a water supply at 1.0 mg/l (P).

It is clear, in general terms, that the environmental impact of dosing ortho-phosphate to a water supply is small. Even so, a local environmental impact assessment should be undertaken as circumstances vary. Generally, there is a need for more research to establish specific ecological objectives and, determine any necessary constraints on phosphate discharges and the need for phosphate elimination in sewage treatment systems. The positive environmental impact of ortho-phosphate dosing will be the avoidance of major disruption, particularly in urban and sub-urban locations.

Key points:

1. *Elevating the pH of water supplies to between 8.0 and 10.0 will much reduce plumbosolvency but may be insufficient for meeting modern day standards for lead in drinking water in some water supply systems.*
2. *Ortho-phosphate is the most effective corrosion inhibitor for reducing plumbosolvency.*
3. *The doses of ortho-phosphate used (generally 0.5 to 2.0 mg/l P) are considered to be entirely safe, given that ortho-phosphate concentrations are many thousands of times higher in many carbonated soft drinks.*
4. *Poly-phosphates can exacerbate plumbosolvency.*
5. *Ortho-phosphate doses are water supply specific and normally range from 0.5 to 2.0 mg/l (P) to achieve effective plumbosolvency control.*
6. *In general, the environmental impact of dosing ortho-phosphate to a water supply system will be small, although specific local circumstances should be considered. In some cases, constraints on dosing ortho-phosphate may apply.*
7. *Ortho-phosphate dosing will avoid major disruptions in Cities and Towns.*

9 Control of materials

9.1 The materials that can leach lead

From the AWWARF report (2008), the following conclusions can be drawn:

(1) Lead service lines: The impact on lead in drinking water at the tap is significant.
(2) On-site lead piping: The impact on lead in drinking water at the tap is significant.
(3) Faucets and immediate connective piping: The impact on lead in drinking water at the tap is less significant.
(4) Water Meters: The impact on lead in drinking water at the tap is insignificant.

In the US, a schedule for the removal of lead service lines owned by the water system is required by the Lead and Copper Rule, when the system continues to exceed the 90th percentile Action Level for successive monitoring periods. However, domestic lead pipes and residence-owned service line segments are mostly only removed at the discretion of the home or building owner. In Europe at the EU level, there is no regulatory requirement to remove lead service pipes nor domestic lead pipes, although regulations in the UK require a water company to remove their lead service line if the home-owner decides to remove their domestic lead pipes.

It is important to note that large numbers of water systems in the U.S., having no lead service lines or interior lead piping, required some form of enhanced corrosion control as a result of the Lead and Copper Rule, due to a combination (not differentiable from the monitoring data) of leaded brasses and soldered joints in copper plumbing. Therefore, the significant lead release observed in the AWWARF study represented lead contamination from waters already adjusted for some corrosion control, and would not represent nearly a worst-case scenario as may be present elsewhere in North America or Europe.

Meters

The AWWARF study reported lead concentrations from exposure to brass in meters that are generally made from brass (copper alloy) materials that contain lead at percentages ranging from 1.5 to 8.0% was up to 110 ppb when exposed to chloraminated water and 58 ppb when exposed to chlorinated water supplies, after 6–7 hours of stagnation. Chloraminated water containing phosphates had an average lead concentration of 17 ppb after 3.5 to 5.5 hours of stagnation.

On-site piping materials

The primary sources of lead onsite include brass faucets, leaded solder, galvanized piping, and other brass fittings. Another study of 25 homes that were at least 20 years old involved collecting a standing 125-mL followed by 1000-mL sample. The average lead levels from these samples were 3.5 ppb (125-mL sample) and 4.3 ppb (1000-mL sample).

Proprietary brasses

To address the concerns of the regulations in California, most of the large brass foundries have begun to utilize proprietary brasses with essentially no lead. Additives such as antimony are substituted into the brass for lead to improve machinability. Two of these proprietary brasses are Federalloy and SeBilloy.

Clearly, if the lead levels are essentially zero, then the contribution of the brass parts will be essentially zero. As noted in the AWWARF (2008) study, with the exception of the faucets, the other brass materials, meters and fittings, have relatively low impact on the lead in first draw samples from the consumer's water system in typical circumstances. However, when manufacturing processes result in segregation of lead at the water contact surface, high lead release can be observed for some indeterminable length of time from a very low lead content (on the average) alloy or device.

9.2 Screening and testing methods

In Germany, automatic pipe rigs have been used which can be constructed with any configuration of pipework and connective fittings. Brass fittings can be coupled with copper piping to investigate lead leaching. These pipe rigs take samples after a range of water stagnation periods and enable the time dependency of metal leaching to be characterised.

In the UK, the lead leaching from brass manifolds was investigated by filling with a range of test waters and measuring lead concentrations after a range of water stagnation periods.

Another simple approach is to immerse the test material, for example, a brass fitting, in an appropriate volume of test water and measure the lead concentration after a period of time.

In North America, many manufacturers of valves, meters, faucets and fittings submit their products for certification with respect to lead and selected other metal leaching characteristics under the NSF/ANSI 61 Standard: either Section 9 for "end point devices" (last 1 litre before the consumer), or Section 8 for "in-line devices".

Key points:

1. *Other than lead pipes, the next most significant source of lead is from brass fittings.*
2. *Lead leaching from brass can be investigated by using pipe rigs or by simple experimentation.*

10 Investigational methods

10.1 Laboratory plumbosolvency testing

Introduction

Taking samples at consumers' taps to determine exposure to lead from drinking water has a number of limitations:

- there can be problems with access;
- lead emissions from taps vary considerably, depending on how long the water has stood in the lead pipes;
- the sampling methods that can be used (see Chapter 3) are either very onerous (e.g.: composite sampling) or prone to sampling error (e.g.: random daytime sampling).

This makes decisions on corrective action difficult. A complimentary approach is to evaluate the plumbosolvency characteristics of the drinking water within a supply area. This can be done using a simple laboratory based method (Colling *et al.*, 1987).

Sampling strategy

The plumbosolvency of a water supply is determined by the quality of the source water(s) and by water temperature. There are exceptions, such as lead leaching from brass and galvanic corrosion effects (see Chapter 1), but case studies indicate that generally the worst lead in drinking water problems relate to the presence of lead pipes. We can therefore focus on the interaction of water with lead pipes.

For a particular water supply area, the first question will be "where does the water come from?" The simplest case will be a single discrete source from a water treatment works, in which case a single sample of the treated water will be appropriate for plumbosolvency testing. In cases where there are two or more source waters, a plumbosolvency testing strategy could be to sample and test separately each of the sources waters, and/or to test representative samples from the distribution system.

For source waters, it is normal practice to sample the final treated water, as pumped into supply. If caustic chemicals are dosed for pH correction, sample after the water has had sufficient contact with the lime, so that it is representative of the water in supply. If corrosion inhibitors are dosed, you can test samples both before and after dosing, to determine the effect of the inhibitor and its dose.

The test requires around 50 litres of water sample and the use of 4×12.5 litre containers is recommended for ease of carrying. Sampling containers should be made of food grade plastic, tightly stoppered and when used kept cool if possible. Samples should not be exposed to direct sunlight, to minimise any algal growth that might occur. Samples should be transported by courier, within two days if possible. Sample containers can be re-used after thorough washing and disinfection with diluted hypochlorite solution. When sampling, ensure sampling lines are fully flushed and also flush out the sampling container with sample water before filling. To undertake laboratory based treatment trials, multiple numbers of 50 litre samples will be required (often, 5×50 litre samples are used to investigate a treatment variable).

Rapid plumbosolvency testing procedure

The testing method that has been used extensively in the UK is based on the work of Colling *et al.* and involves pumping the 50 litre sample of water through duplicate (in parallel) 15 cm lengths of new 12 mm internal diameter lead piping, at a flow rate (approximately 0.5 ml/min) to provide 30 minutes contact (30MC) between the test water and the piping. Each section of lead piping, sealed with rubber bungs, is immersed in a water bath and maintained at a constant 25°C for a test period of typically 20 days. This configuration is shown in Figure 10.1.

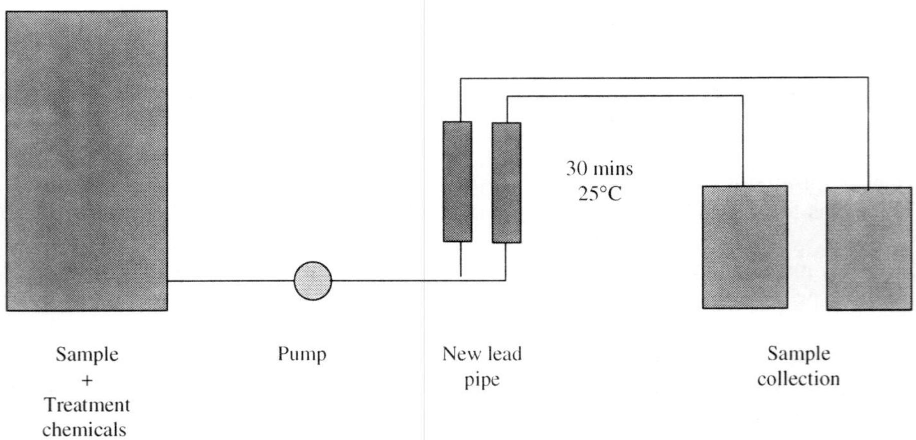

Figure 10.1 Configuration of plumbosolvency testing equipment

During the test period, the flow through each section of lead piping is checked three times a week, at the same time that test water pH is checked and when a small sample of the test water leaving the lead pipe section is taken for lead analysis. At these times, pH adjustments are made if necessary using small additions of acid or alkali. At the end of the test period, the test water is allowed to stagnate for 16 hours within the lead pipe section, prior to a further sample being obtained for lead analysis.

Lead should be analysed using an appropriate method, which must be subject to routine quality control procedures.

A 32-channel peristaltic pump can be used, enabling 15 duplicate test samples and a duplicate control sample (e.g.: a stable local groundwater derived tap-water) to be tested simultaneously. This allows various experimental designs to be used in investigating the plumbosolvency characteristics of test waters, examples being:

(i) 15 different test waters as part of a regional survey, or
(ii) three test waters, each at a particular pH, for five levels of ortho-phosphate addition, or
(iii) a single test water over three pH conditions for five levels of ortho-phosphate addition.

The latter more exhaustive experimental design for a single test water has been used commonly in the UK for investigating the treatment responses of water supply schemes. The lead pipe sections used in this testing were manufactured by the same foundry, their 5 metre length mouldings being cut into 15 cm lengths prior to delivery, thereby avoiding pipe bending (anecdotal evidence had earlier suggested that pipe bending had caused variation in water lead emission characteristics). In this work, samples were either delivered to the laboratory on the day of sampling or put on an over-night courier. Testing commenced normally within a day of sample receipt and samples were kept at ambient room temperature but covered by black plastic sheeting to exclude light. The reproducibility and pattern of results obtained over several years gave no indication of any problems arising from sample handling, other than slight pH drift which was easily corrected during testing.

The typical results obtained for high alkalinity test waters, with and without ortho-phosphate, are shown in Figure 10.2. Some variation in the lead emissions can often be apparent over the test period, particularly for test waters without ortho-phosphate, and for comparative purposes the data can be summarised as follows:

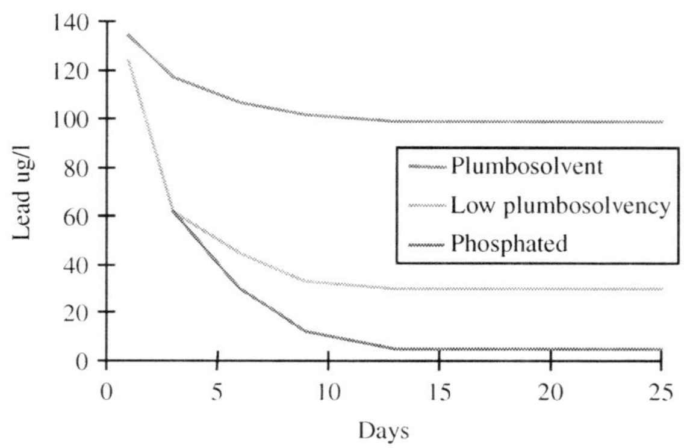

Figure 10.2 Lead emissions from pipe sections at 25°C after 30 minutes contact for high alkalinity waters

(i) lead results from the first week of testing are ignored as this is commonly the period when the test water is found to equilibrate with the lead piping;
(ii) the median lead concentration of the subsequent lead emissions from each section of lead piping can then be determined – a comparison of the median lead concentrations from each related pair of lead pipe sections is the principal means of checking reproducibility of the test procedure;
(iii) the average lead emission after 16 hours stagnation from the duplicate pair of lead pipe sections is then determined.

In general, the test procedure should have the following performance characteristics:

(i) the test temperature remains constant within ± 0.1°C
(ii) the flow rate is stable within ± 10%
(iii) an average contact time of 30 minutes is achieved within ± 1 minute
(iv) pH is maintained within ± 0.2 of target
(v) the analytical limit of detection for lead is 3 μg/l or lower as this can readily differentiate a wide range of test waters, in which median emissions might range from this detection limit up to several thousand μg/l.

The only significant problem that may be experienced in testing can occur with iron-rich, organically coloured upland surface waters, when plating out of iron precipitates can cause occasional blockage of the peristaltic tubing. Work by Colling *et al.* (1992) indicated that the lead emissions in equilibrium with sections of new lead pipe were similar to sections of old exhumed lead pipe, except that in the latter case the time taken for equilibrium to become established was much longer (months as opposed to days with new lead pipe). They also found that lead emission concentrations approximately halved if the test temperature was reduced from 25 to 12°C. Figure 10.3 shows a typical set of summary results from which the relationships between lead emissions and treatment conditions can be readily appreciated.

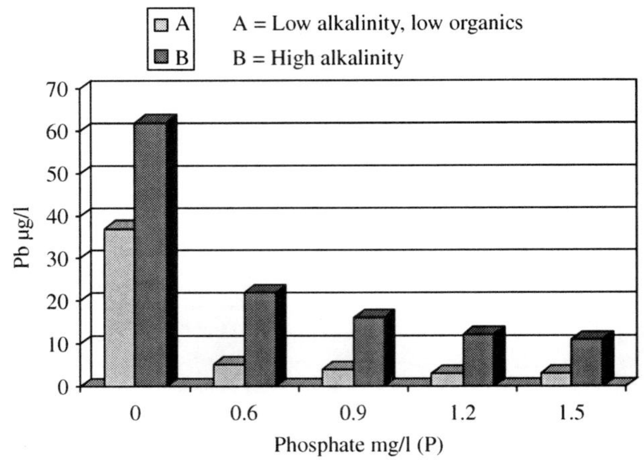

Figure 10.3 Averaged median lead pick-up under a range of ortho-phosphate dosing conditions at 25°C

10.2 Lead pipe test rigs

Lead pipe test rigs have many possible designs, including closed-loop and pass-through, the latter being more common. The lead pipe can be a straight length, a coil or swan-necked. In the UK, the swan-necked design has been the most popular and an automated example is shown in Figure 10.4. Similar approaches have been used in the US (AWWARF, 1990; Cantor, 2009).

In the lead pipe test rig shown, the operating principles are:

(1) Water is flushed to waste.
(2) Water is then held for a stagnation period of 30 minutes (or any other period).
(3) The stagnated water is then captured by the sampling vessel which displaces earlier sample water to waste.
(4) The cycle is then repeated so that a fresh stagnation sample is available to be picked up at any time.

New lead pipes were mostly used in the UK because of the erratic lead emissions that often derive from exhumed lead pipes.

The typical deployment (see also Section 8.4) provides:

(1) Treated water at the treatment works, before and after ortho-phosphate dosing: a direct measure of the impact of the ortho-phosphate dosing.
(2) Within the near distribution system, close to the treatment works: to demonstrate the plumbosolvency of freshly distributed water.
(3) Within the far distribution system, distant from the treatment works: to demonstrate any effects from older water (such as iron discolouration).

Figures 10.5 and 10.6 give illustrative results from test rigs using new lead pipe.

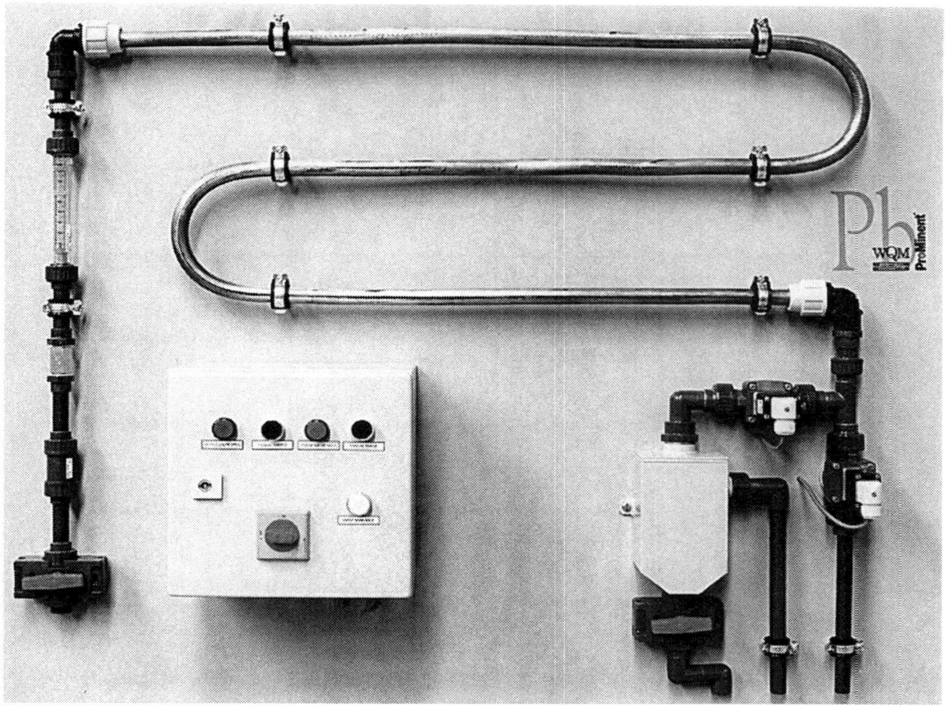

Figure 10.4 Example of a lead pipe test rig

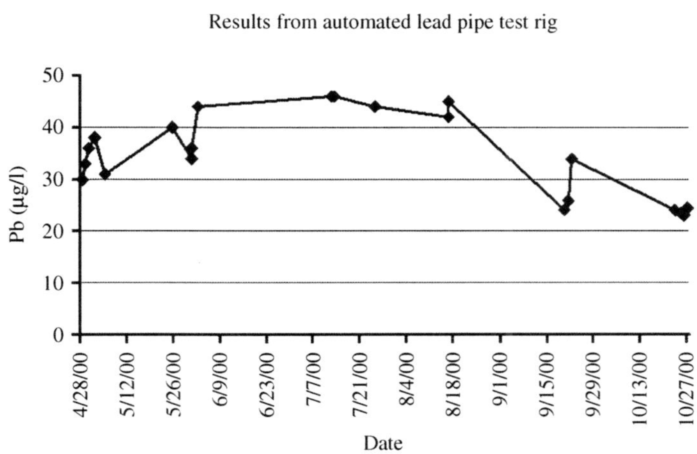

Figure 10.5 Test rig results for a treated water in Western England before ortho-phosphate dosing

10.3 Zonal compliance modelling

Introduction

As discussed in Chapter 3, physical sampling of lead emissions in a water supply zone has limitations. The most appropriate method for establishing average lead concentrations at a consumer's tap is split-flow

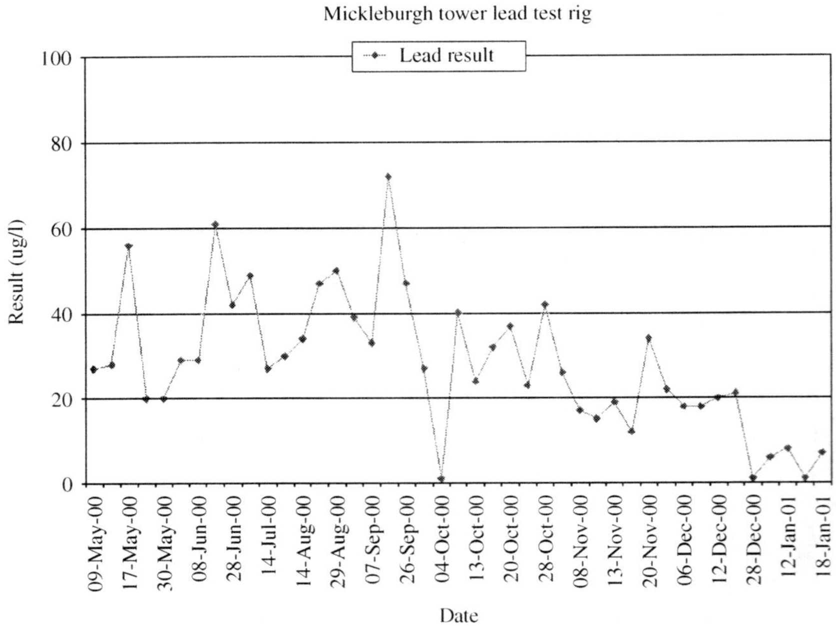

Figure 10.6 Test rig results at a water tower in Eastern England with variable mixing of two sources of different plumbosolvency

composite sampling, as used by Van den Hoven *et al.* (1999), but this is much constrained by logistics and cost. Random daytime sampling is much easier to implement at lower cost, but reproducibility is poor unless high numbers of samples are taken. This raises a number of questions:

- How can the extent of lead problems in a zone be determined?
- What will be the likely outcome of any intervention measures?
- How confident can we be that we have achieved compliance with the regulatory standards?

Part of the answer to these questions is to use computational modelling, in association with supporting techniques.

How lead emissions across a zone can be predicted

The models, which are described in more detail elsewhere (Van der Leer *et al.*, 2002), enable the most relevant features of a water supply zone to be incorporated in the prediction of zonal compliance with lead standards, as a function of both plumbosolvency (corrosivity of the water to lead) and the zone's physical characteristics. A zonal model simulates the emissions of lead at individual simulated houses, through time, across an entire water supply zone or area of supply. It uses a single pipe model to determine the lead emission profile at each simulated house, the characteristics of each simulated house being the outcome of the random ascription of variables, which follows the Monte Carlo method for establishing a probabilistic frame-work.

The single pipe model simulates the dissolution of lead into the water flowing through or stagnating in a coupled lead pipe and non-lead pipe, over a 24-hour period (particulate lead is not simulated). The coupled pipes are broken down into a series of "elements" and when assuming simple plug flow, each element is treated as a stirred tank, flow being simulated by passing the contents of each stirred tank to the next at a time interval of one second. The rate of lead dissolution is determined by reference to an exponential curve that declines towards equilibrium, as illustrated in Figure 3.1. The model does not simulate lead release from brass or galvanic corrosion, nor particulate lead releases.

As M (the initial mass transfer rate which determines the initial slope of the dissolution curve) and E (the equilibrium concentration) reduce, the water is less plumbosolvent (less lead dissolves: curves A to C) and these factors can be determined by stagnation sampling at appropriate reference houses or by laboratory plumbosolvency testing. Curves A1 and A2 differ in shape as a consequence of the relationship between the 30 minutes stagnation and equilibrium concentrations, which vary for individual waters (Hayes, 2008). The exponential curve and the assumption of plug flow are both approximations, but they enable the computational demands of the model to be greatly reduced. Extensive research (Hayes, 2002 and Van der Leer *et al.*, 2002) has demonstrated that these approximations are adequate when compared to the more scientifically exact diffusion model and the three dimensional simulation of turbulent flow.

As a guide, a moderately plumbosolvent water (e.g.: a high alkalinity groundwater that has not been phosphated, or a low alkalinity water that has been pH adjusted to 8.0–8.5 but not phosphate dosed) will often be described by $M = 0.1$ and $E = 150$. For 12 mm internal diameter lead pipes, the equilibrium concentration (E) is predicted (Hayes, 2002) to occur after about 8 hours water stagnation, consistent with the observations and predictions of others (Kuch and Wagner, 1983).

When the imaginary tap is closed (that is, zero flow), the lead concentration increases over time as determined by M and E. When the imaginary tap is open, the concentration of lead in the emission from the pipe is either (i) a reflection of the steady state flushed condition (with lead concentrations normally below 1 µg/l unless the lead pipe is very long) or (ii) it is determined by previous zero flow (stagnation) conditions, as influenced by pipe geometry and the extent of the flow event. It can be appreciated that the simulation of such events in each stirred tank for every second of flow leads to millions of calculations being performed for each simulated pipe.

The zonal model is set up by the random ascription of a series of zonal characteristics, as derived from sets of agreed statistical distributions, and by the use of agreed variables and constants. The statistical distributions that are shown in Figure 10.7 have been used successfully in many zonal modelling studies (e.g.: Hayes *et al.*, 2008). They have the following features:

- the length of lead and non-lead pipes have a log-normal distribution, consistent with longer lengths occurring less frequently;
- for the lead pipes, 95% are assumed to have an internal diameter of 12mm and 5% 18mm, as relates to UK conditions;
- the volume used per day relates to an individual simulated house, the mean volume equating to the average water consumption of a house in the UK and assumed to flow through the simulated pipes;
- pattern A describes water usage in a house in which there is residency throughout the hours when water is consumed (not during the night when residents are asleep);
- pattern B describes water usage in a house in which all residents are absent during "office hours" when no water is used;
- patterns A and B are applied for three and two water use frequencies, such that the weighting of A to B is 3 to 2, albeit with the water use frequencies within the two categories having an equal weighting.

It is easy to change these statistical distributions if local survey data is available.

The aim of this probabilistic Monte Carlo framework is to describe the huge variation that undoubtedly occurs in real water supply zones. If we can mimic this real-world variation then the model can be used for predictive purposes, as has been demonstrated by case studies (Hayes *et al.*, 2006, 2008). It should be appreciated that the average lead concentrations predicted by the model relate to a single plumbosolvency condition occurring in time, whether it is applied as a constant throughout an area or as a range. In consequence, the predicted results relate to an average condition over time. This is reasonable if the periods of time under consideration extend to multiples of a year, such that seasonal variation is accommodated.

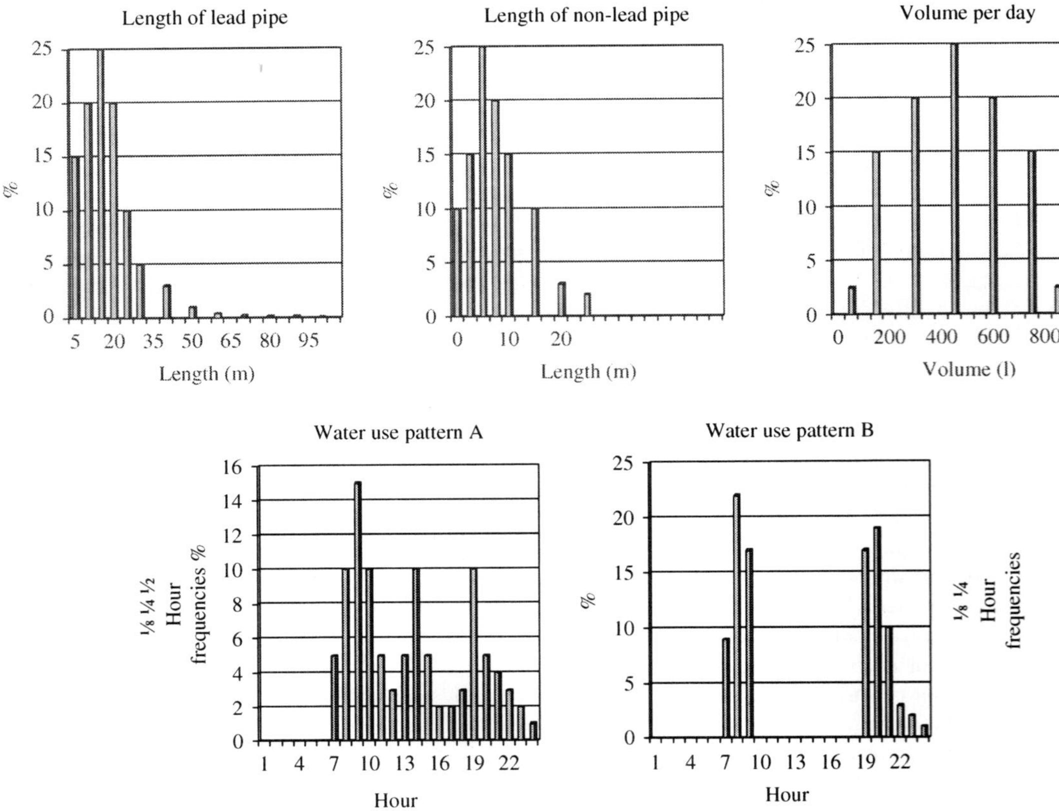

Figure 10.7 Statistical distributions used to set up the zonal model (from Hayes, 2009)

The model, as used in numerous UK case studies, simply assumes that a single lead pipe supplies each simulated house, as illustrated in this chapter. However, the model does have an additional optional feature in which branched lead pipes can be simulated, to accommodate better multiple dwelling buildings (e.g.: apartments), albeit case studies are presently limited.

The zonal model calculates the daily average concentration (DAC) of the lead emissions for each simulated house and from this can readily determine the percentage of simulated houses that fail a series of specified standards (typically: 10, 25 and 50 μg/l). As the zonal model uses a range of water use patterns (Figure 10.7) that also span week-day and week-end consumptions, the DAC is taken to be equivalent to the weekly average concentration, and equivalent to composite sampling (COMP). This is of interest as the EU directive (European Commission, 1998) describes the lead standards in terms of weekly average lead concentrations ingested.

It is of course not possible to validate these DAC outputs directly without exhaustive composite sampling (which is not logistically feasible) and so a sampling model is used, in order to characterise the behaviour of the simulated zone in a way that can be validated by the data collected by water companies. Random daytime (RDT) sampling is of greatest relevance in the UK, as it has been used for regulatory purposes for many years (UK Government, 1989).

In order to simulate a RDT survey, the specified number of simulated houses are selected at random and then a sampling time is selected at random between the hours of 09-00 and 17-00. The RDT sample is simulated by a stirred tank of one litre capacity as the outlet from the pipe. At the time of simulated sampling, the pattern of water use that has been applied to the simulated house is used to determine the immediately previous water – pipe contact position. It is routine to repeat the simulated survey, typically

100 times, in order to be able to understand possible variation. The result reported for the zone under investigation is the average survey result from all the surveys simulated, although confidence bands are also computed and can be used if required in model validation. Examples of the validation of the zonal modelling procedure are given in Table 10.1 for pre- and post- orthophosphate dosing conditions.

Table 10.1 Validation examples: predicted and observed zonal failure rates for RDT samples (from Hayes *et al.*, 2006, 2008)

Area and status	Predicted and observed RDT samples	% Failing 10 μg/l	% Failing 25 μg/l	% Failing 50 μg/l
Cambridge – before phosphate	Predicted	28.4	15.1	5.2
	Observed	32.4	15.0	4.5
Cambridge – after phosphate	Predicted	0.7	0.0	0.0
	Observed	0.5	0.0	0.0
South East Wales – before phosphate	Predicted	18.4	11.3	4.8
	Observed	21.8	11.8	4.5
South East Wales – after phosphate	Predicted	<1.9	<0.1	0.0
	Observed	1.7	0.9	0.9

Table 10.2 shows how the model can be used to predict compliance with standards for different assumed reductions in plumbosolvency, as can be achieved using a corrosion inhibitor. This is shown for RDT sampling but other sampling methods can be predicted.

Table 10.2 Use of zonal model to predict RDT sampling compliance for a range of plumbosolvency reductions when 50% houses have a lead pipe

% Pb reduction	M (μg/m^2/sec)	E (μg/l)	% Houses >10 μg/l	% Houses >25 μg/l	% Houses >50 μg/l
0	0.100	150	27.57	10.97	2.68
60	0.040	60	10.90	1.50	0.02
70	0.030	45	6.27	0.51	0.00
80	0.020	30	2.61	<0.01	0.00
90	0.010	15	0.18	0.00	0.00

In this example, it can be interpolated that an 82% reduction in plumbosolvency will be needed to achieve a 98% compliance level with the 10 μg/l standard, the optimisation target in the UK. The model can also be used to predict the outcome of a partial lead pipe replacement campaign. Illustrative results are shown in Table 10.3. An overview of how the model is used is given in Figure 10.8.

Table 10.3 Prediction of the results of partial lead pipe replacement for a zone with 50% houses with a lead pipe and moderately plumbosolvent water (M = 0.1, E = 150)

Lead pipes removed	% >10 μg/l	% >25 μg/l	% >50 μg/l
None	20.19	8.05	2.04
All 5m (15%)	18.17	7.87	2.06
All 5 & 10 m (35%)	15.34	7.16	1.84
All 5, 10 & 15 m (60%)	10.65	5.72	1.79

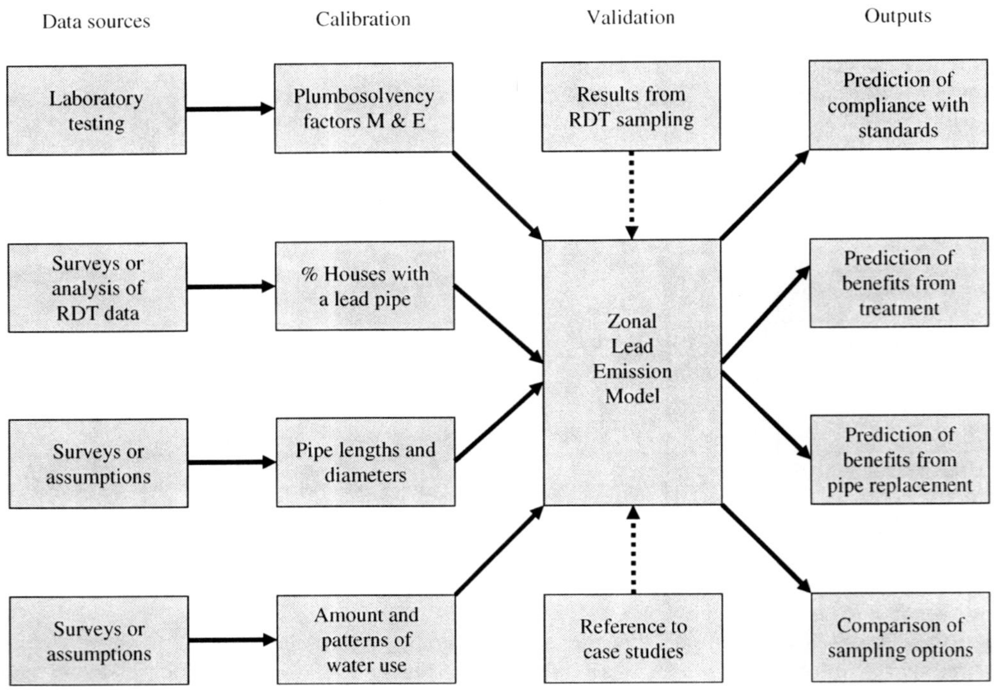

Figure 10.8 Calibration, validation and use of the zonal model

Key points:

1. *Laboratory testing using sections of new lead pipe is reproducible and can determine the plumbosolvency of drinking water.*
2. *Test results are available within one month.*
3. *The impact of pH changes on plumbosolvency can be quickly investigated.*
4. *The test can determine which corrosion inhibitor works best and its dose response.*
5. *Computational modelling can predict lead emissions across an entire water supply zone,*
6. *The modelling system has been validated in UK case studies.*
7. *The model can predict the results of dosing a corrosion inhibitor and assist in dose optimisation.*
8. *The model can predict the results of partial lead pipe replacement (where there is no galvanic corrosion component).*

11 Economics of plumbosolvency control

11.1 Ortho-phosphate dosing

Unit treatment costs

At August 2009, the cost of ortho-phosphoric acid (75% w/w H_3PO_4) in the UK was £750 (GBP) per tonne. At a dose of 1.0 mg/l (as P), this equates to a unit chemical cost of 0.32 p/m^3 (£3.20 per Ml). Allowing for related operational costs (power, maintenance, monitoring, etc.) at 0.13 p/m^3 (£1.30 per Ml) the total operational cost of dosing 1.0 mg/l (as P) is 0.45 p/m^3 (£4.50 per Ml). In Euros, assuming £1.00 = €1.16, the total operational cost of dosing 1.0 mg/l (as P) is 0.52 c/m^3 (€5.20 per Ml).

The unit chemical cost will rise or fall depending on the market price of ortho-phosphoric acid, which fluctuates. The unit chemical cost will also be dose dependent. Operational over-heads are not dependent on the cost of ortho-phosphoric acid and will change only slightly with dose. Operational over-heads will vary from country to country, particularly labour costs.

Capital cost of dosing plant

Small scale dosing plants will cost from €20,000 to €50,000 assuming kiosk housing, a flow proportional dosing pump, the use of an intermediate bulk container for acid storage and attendant services (power, dosing lines, water, drainage). Larger scale dosing plants will cost from €100,000 to €200,000 assuming bulk acid storage.

Cost example

For a water supply system serving 100,000 people, an average water consumption of 150 litres per head per day and 25% leakage in the distribution network, the total volume to be treated will be 20,000 m^3/d (20 Mld). For a dose of 1.0 mg/l (as P), the treatment cost will be €37,960 per annum.

11.2 Lead pipe replacement

Unit costs of replacing lead service pipes

Units costs are likely to range from around €500 to €2,000 per lead service pipe replaced, depending on local labour costs and ground conditions (particularly the extent of paved and road surfaces). In Brussels (Belgium) over the period 2002 to 2008, 38,000 lead connections (owned by the water company) were replaced at a total cost of €15M, averaging €395 per connection. In the Hague

(Netherlands) over the period 1985 to 2004, over 97,000 lead connections were replaced at an average cost of €564 per connection.

Unit costs of replacing domestic lead pipes

Units costs are likely to range from €500 to €2,000 per domestic lead pipe replaced, depending on pipe lengths, ground conditions, re-instatement of garden frontages and internal re-decoration.

Unit costs of total lead pipe replacement

Units costs are therefore likely to range from €1,000 to €4,000 for total lead pipe replacement per property.

Cost example for total lead pipe replacement

For a population of 100,000 and an average residency of 2.5 people per dwelling, the water supply system will serve 40,000 dwellings. If 50% of these dwellings (i.e., 20,000) have a lead service pipe, the replacement cost will be between €10 M and €40 M. If 25% of these dwellings (i.e., 10,000) have a domestic lead pipe, the replacement cost will be between €5 M and €20 M.

Therefore, the cost of total lead pipe replacement will be between €15 M and €60 M.

Unit costs of other options

Epoxy lining is likely to cost between €500 and €2,000 per connection, whereas internal cathodic protection is likely to cost between €250 and €1,000 per connection depending upon the nature of paved surfaces and cleaning method.

11.3 Cost comparisons using net present value

Based on a 25 year planning horizon, the costs identified in Sections 11.1 and 11.2 for a City of 100,000 population, and a discount rate of 6%, the following NPVs can be calculated:

[1] Treatment assuming €100,000 capital cost, 20% capital replacement every 5 years and annual operating costs of €37,960

NPV = €0.655 M

[2] Total lead pipe replacement assuming a cost of €15 M spread over 10 years

NPV = €11.668 M

[3] Total lead pipe replacement assuming a cost of €60 M spread over 10 years

NPV = €46.754 M

Key point:

1. *Dosing ortho-phosphate to control plumbosolvency costs considerable less than lead pipe replacement.*

12 References

Agency for Toxic Substances and Disease Registry (2007). Toxicological profile for lead. US Department of Health and Human Services.

American Waterworks Association (1990). Lead service line replacement: Benefit to cost analysis. Water Industry Technical Action Fund.

American Waterworks Association Research Foundation (AWWARF) (1990). Lead control strategies. Lead Control Strategies. AWWA Research Foundation and AWWA, Denver, CO.

American Waterworks Association Research Foundation (AWWARF) (2008). Contribution of Service Line and Plumbing Fixtures to Lead and Copper Rule Compliance Issues.

Assennato, G., Paci, C. and Baser, M.E. (1987). Sperm count suppression without endocrine dysfunction in lead exposed men. *Archives of Environmental Health*, **42**, 121–127.

Baron, J. (2001). Monitoring strategy for lead in drinking water at consumer's tap: field experiments in France: Water Science and Technology. *Water Supply*, **1**(4), 193–200.

Beattie, A.D., Moore, M.R., Devenay, W.T., Miller, W.T. and Goldberg, A. (1972). Environmental lead pollution in an urban soft-water area. *British Medical Journal*, **1**, 491–493.

Bellinger, D.C. (2008). Very low lead exposures and children's neurodevelopment. *Current opinion in Paediatrics*, **20**, 172–177.

Boyd, G.R., Shetty, P., Sandvig, A.M. and Pierson, G.I. (2004). Pb in tap water following simulated partial lead pipe replacement. *Journal of Environmental Engineering. American Society of Civil Engineers*, Washington DC, **130**, 1188.

Breach, R., Crymbal, S. and Porter, M.J. (1991). Systematic approach to minimising lead levels at consumers' taps. *Proceedings of the AWWA Annual Conference.*

Britton, A. and Richards, W.N. (1981). Factors influencing plumbosolvency in Scotland. *Journal of the Institue of Water Engineers and Scientists*, **35**(5), 349–364.

Canfield, R.L., Kreher, D.A., Cornwell, C. and Henderson Jr., C.R. (2003). Low-level lead exposure, executive functioning, and learning in early childhood. *Child Neuropsychology*, **9**, 35–53.

Cantor, A.F. (2009). Water distribution system monitoring. CRC Press.

Cardew, P.T. (2009). The impact of ortho-phosphate treatment on lead in drinking water in North West of England. *Proceedings of International Conference*, COST Action 637, October 2008, Lisbon.

CDC (1991). Preventing lead poisoning in young children. US Department of Health and Human Services, Public Health Service, Atlanta, Georgia.

CDC (2005). 3rd National report on human exposure to environmental chemicals. NCEH publication no. 05-0570 http://www.cdc.gov/exposurereport/

CDC (2010). Important update: lead-based water lines. www.cdc.gov/nceh/lead/waterlines.htm

Chiodo, L.M., Covington, C., Sokol, R.J., Hannigan, J. H., Jannise, J., Ager, J., Greenwald, M. and Delaney-Black, V. (2007). Blood lead levels and specific attention effects in young children. *Neurotoxicology and Teratology*, **29**, 538–546.

Colling, J. H., Whincup, P.A.E. and Hayes, C.R. (1987). The measurement of plumbosolvency propensity to guide the control of lead in tapwaters. *Journal of the Institution of Water and Environmental Management*, **1**(3), December 1987.

Colling, J.H., Croll, B.T., Whincup, P.A.E. and Harward, C. (1992). Plumbosolvency effects and control in hard waters. *JIWEM*, **6**, 259–268, June 1992.

COST Action 637 (2008). Research Seminar, April 2008, Brussels.

Croll, B.T. (2000). Applying science to the optimisation of plumbosolvency control. *Proceedings of a Technical Seminar on Lead in Drinking Water*, 3–8, CIWEM, London, 1st December 2000.

Dervany, D. (2009). COST Action 637, Research Seminar, April 2009, Budapest.

Dietrich, K.N., Krafft, K.M. and Bornschein, R.L. (1987). Low-level foetal lead exposure effect on neurobehavioural development in early infancy. *Pediatrics*, **80**, 721–730.

Douglas, I., Guthmann, J., Muylwyk, Q. and Snoeyink, V. (2004). Corrosion control in the City of Ottawa-Comparison of alternatives and case study for lead reduction in drinking water. In: W. Robertson and T. Brooks (eds.), 11th Canadian National Drinking Water Conference and 2nd Policy Forum, April 3–6, Calgary, AB. Canadian Water and Wastewater Association, Ottawa, ON.

Douglas, I., Campbell, A. and Muylwyk, Q. (2007a). Lead control in Soft Water: Experience from a Canadian Utility. *Proc. AWWA Annual Conference*. Toronto, ON. American Water Works Association, Denver, CO.

Douglas, I., Lemieux, F. and Weir, D. (2007b). Testing the waters: putting Canada's new corrosion control guideline into practice. In: *Proceedings of the 2007 AWWA Research Symposium-Distribution Systems*: The Next Frontier, March 2–3, 2007, Reno, NV. American Water Works Association, Denver, CO.

Drinking Water Inspectorate (2000). Determination of requirements to meet new lead standards. Information Letter 12/2000.

Drinking Water Inspectorate (2001). Further guidance on requirements to meet new lead standards. Information Letter 3/2001.

Edwards, M., Triantafyllidou, S. and Best, D. (2004). Elevated blood levels in young children due to lead-contaminated drinking water: Washington DC, 2001–2004. *Environment Science and Technology*.

Edwards, M. and Triantafyllidou, S. (2007). Chloride-to-sulfate mass ratio and lead leaching to water. *J. Am. Water Works Assoc.*, **99**(7), 96–109.

European Commission (1998). Council Directive (98/83/EC) of 3 November 1998 on the quality of water intended for human consumption. Official Journal, L330/32, 5 December 1998.

European Commission (2008). The quality of drinking water in the European Union. Synthesis report on the quality of drinking water in the Member States of the European Union in the period 1999–2001 Directive 80/778/EEC, 14 April 2008.

Fertmann, R., Hentschel, S., Dengler, D., Janßen, U., Lommel, A. (2003). Lead exposure by drinking water: an epidemiological study in Hamburg, Germany. *International Journal of Hygiene and Environmental Health*, **207**, 235–244.

Gilbert, S.G. and Weiss, B. (2006). A rationale for lowering the blood lead action level from 10 to 2 μg/dl. *NeuroToxicology*, **27**, 693–701.

Hall, A. (1905). The increasing use of lead as an abortifacient. *The British Medical Journal*, 18 March 1905.

Hayes, C.R. (2002). PhD Thesis. Computational modelling of lead in drinking water. University of Wales.

Hayes, C.R. (2008). Optimisation tools for achieving the lead standard of 10 μg/l in drinking water. *Proceedings of an International Conference on Metals and Related Substances in Drinking Water*. October 2007. Antalya, Turkey. COST Action 637.

Hayes, C.R. (2009). Computational modelling to investigate the sampling of lead in drinking water. *Water Research* **43**, 2647–2656.

Hayes, C.R. (2010). Computational modelling methods for assessing the risks from lead in drinking water. *Journal of Water and Health*. In press.

Hayes, C.R., Bates, A.J., Jones, L., Cuthill, A.D., Van der Leer, D. and Weatherill, N.P. (2006). Optimisation of plumbosolvency control using a computational model. *Water and Environment Journal*, **20**, 256–264.

Hayes, C.R., Incledion, S. and Balch, M. (2008). Experience in Wales (UK) of the optimisation of ortho-phosphate dosing for controlling lead in drinking water. *Journal of Water and Health*, **6**(2), 177–185.

Hayes, C.R. and Skubala, N.D. (2009a). A review of lead in drinking water and its impact on health. *Proceedings of the 2nd International Conference on Metals and Related Substances in Drinking Water*. October 2008, Lisbon, COST Action 637.

Hayes, C.R. and Skubala, N.D. (2009b). Is there still a problem with lead in drinking water in the European Union? *Journal of Water and Health*, **7**(4), 569–580.

Health Canada (2009). Guidance on controlling corrosion in drinking water distribution systems. ISBN 978-1-100-14193-0.

Hepple, S. (2009). MRes thesis, Swansea University.

Hoekstra, E.J., Aertgeerts, R., Bonadonna, L., Cortvriend, J., Drury, D., Goossens, R., Jiggins, P., Lucentini, L., Mendel, B., Rasmussen, S., Tsvetanova, Z., Versteegh, A. and Weil, M. (2008). The advice of the Ad-Hoc Working Group on Sampling and Monitoring to the Standing Committee on Drinking Water concerning sampling and monitoring for the revision of the Council Directive 98/83/EC. Office for Official Publications of the European Communities, Luxembourg, EUR 23374 EN – 2008. http://circa.europa.eu/Public/irc/env/drinking_water_rev/library?l = /working_monitoring&vm = detailed&sb = Title

Hoekstra, E.J., Hayes, C.R., Aertgeerts, R., Becker, A., Jung, M., Postawa, A., Russell, L. and Witczak, S. (2009). Guidance on sampling and monitoring for lead in drinking water. Office for Official Publications of the European Communities, Luxembourg, EUR 23812 EN – 2009. http://www.meteau.org/publications/index.php

Hulsmann, A.D. (1991). Particulate lead in water supples. *Journal of the Institue of Water Engineers and Scientists*, **4**(2), 19–25.

International Water Association (2004). Bonn Charter.

Jones, R.R. (1989). The continuing hazard of lead in drinking water. *Lancet*, **8864**, 669–670.

Jung, M. and Heiss, G. (2007). Geogenic and non-geogenic metals in Austrian tap water. *Proceedings of the International Conference on Metals and Related Substances in Drinking Water (METEAU)*. Antalya, Turkey. 29–31 October 2007. COST Action 637.

Kuch, A. and Wagner, I. (1983). A mass transfer model to describe lead concentrations in drinking water. *Water Research*, **17**(10), 1303–1307.

Lacey, R.F., Moore, M.R. and Richards, W.N. (1985). Lead in water, infant diet and blood the Glasgow duplicate diet study. *Science of the Total Environment*, **41**, 235–257.

Lacey, R.F. and Jolly, P.K. (1986). Sampling for household lead. TR244, Water Research Centre.

Macintyre, C., Fulton, M., Hepburn, W., Yang, S., Raab, G., Davis, S., Heap, M., Halls, D. and Fell, G. (1998). Changes in blood lead and water lead in Edinburgh: an eight year follow-up to the Edinburgh lead study. *Environmental Geochemistry and Health*, **20**, 157–167.

Mahaffey, K.R., Rosen, J.F., Chesney, R.W. (1982). Association between age, blood lead concentration and serum 1,25-dihydroxycholecalciferol levels in children. *A. J Clin Nutr*, **35**, 1327–1331

Moore, M.R. (1977). Lead in drinking water in soft water areas-health hazards. *Science of the Total Environment*, **7**, 109–115.

Moore, M.R., Goldberg, A., Meredith, P.A., Lees, R., Low, R.A. and Pocock, S.J. (1979). The contribution of drinking water lead to maternal blood lead concentrations, *Clinica et Chimica Acta*, **95**, 129–133.

Moore, M.R., Robertson, S.J., Gilmour, W.H., Murray, G.D., Britton, A., Low, R.A. and Watt, G.C.M. (1998). Decline of maternal blood lead concentrations in Glasgow. *Journal of Epidemiology and Community Health*, **52**, 672–673.

Němcová, V. (2008). Personal communication.

Ontario Ministry of the Environment (2007). http://www.e-laws.gov.on.ca/html/regs/english/elaws_regs_030170_e.htm#BK151

Pocock, S.J., Smith, M., Baghurst, P. (1994). Environmental lead and children's intelligence- a systematic review of the epidemiologic evidence. *British Medical Journal*, **309**, 1189–1197

Quinn, M.J. and Sherlock, J.C. (1990). The correspondence between U.K. "action levels" for lead in blood and in water. *Food Additives and Contaminates*, **7**, 387–424.

Ransom, W.B. (1900). On lead encephalopathy and the use of diachylon as an abortifacient. *The British Medical Journal*, 30 June 1900.

Renner, R. (2006). Lead in water linked to coagulant. *Environ. Sci. Technol.*, **40**(17), 5164–5165.

Ruebel and Becker (2008). Personal communication.

Schock, M.R. (1989). Understanding corrosion control strategies for lead. *Journal of the American Water Works Association*, **81**, 88–100.

Schock, M.R. and Lytle, D.A. (2010). Internal Corrosion and Deposition Control. Water Quality and Treatment: A Handbook of Community Water Supplies. McGraw-Hill, Inc., New York. (In press).

Schock, M.R., DeSantis, M.K., Metz, D.H., Welch, M.M., Hyland, R.N. and Nadagouda, M.N. (2008). Revisiting the pH Effect on the Orthophosphate Control of Plumbosolvency. *Proc. AWWA Annual Conference*, Atlanta, GA.

Schock, M.R., Wagner, I. and Oliphant, R. (1996). The Corrosion and Solubility of Lead in Drinking Water. *Internal Corrosion of Water Distribution Systems*. AWWA Research Foundation/DVGW Forschungsstelle, Denver, CO, pp. 131–230.

Schwartz, J. and Otto, D. (1991). Lead and minor hearing impairment. *Archives of Environmental Health*, **46**, 300–305.

Skubala, N.D. and Hayes, C.R. (2009). A review of lead, copper and nickel in European drinking water. *Proceedings of the 2nd International Conference on Metals and Related Substances in Drinking Water*. October 2008, Lisbon, COST Action 637.

Scottish Executive (2007). Letter to Scottish Water, 23 February 2007.

Slaats, N. (2008). Personal communication.

Tong, S.L., Baghursr, P., McMichael, A., Sawyer, M. and Mudge, J. (1996). Lifetime exposure to environmental lead and children's intelligence at 11–13 years: The Port Pirie cohort study. *British Medical Journal*, **312**, 1569–1575.

Triantafyllidou, S., Nguyen, C. and Edwards, M. (2009). Contribution of galvanic corrosion to lead (Pb) in water after partial lead service line replacement. WQTC '09.

Troesken, W. (2006). The great lead water pipe disaster. MIT Press.

UK Government (1989). Guidance on safeguarding the quality of public water supplies. Department of the Environment. HMSO. ISBN 0 11 7522627.

United Nations Economic Commission for Europe/WHO (2007) Protocol on Water and Health. GE.06-26870-January 2007-4.290. Geneva.

United Nations (2009). Guidance on setting targets, evaluation of progress and reporting. ECE/MP.WH/WG.1/2009/4, EUR/08/5086340/9.

Van den Hoven, T.J.L., Buijs, P.J., Jackson, P.J., Gardner, M., Leroy, P., Baron, J., Boireau, A., Cordonnier, J., Wagner, I., do Mone, H.M., Benoliel, M.J., Papadopoulos, I. and Quevauviller, P. (1999). Developing a new protocol for the monitoring of lead in drinking water. EUR 19087.

Van der Leer, D., Weatherill, N.P., Sharp, R.J. and Hayes, C.R. (2002). Modelling the diffusion of lead into drinking water. *Applied Mathematical Modelling*, **26**(6), 681–699.

Van Dongen, C., Oorthuizen, W. and Tielemans, M. (2008). Central conditioning and lead pipe removal in the drinking water supply of the Hague. *Proceedings of an International Conference on Metals and Related Substances in Drinking Water. Antalya*, Turkey. 29–31 October 2007. COST Action 637.

Veschetti, R. (2008). Personal communication.

Watt, G.C.M., Britton, A., Gilmour, H.G., Moore, M.R., Murray, G.D. and Robertson, S.J. (2000). Public health implications of new guidelines for lead in drinking water: a case study in an area with historically high water lead levels. *Food and Chemical Toxicology*, **38**, S73–S79.

Witczak and Postawa (2008). Personal communication.

World Health Organization (2004). Guidelines for Drinking-water Quality: Third Edition, Vol. 1, Recommendations, WHO Geneva.

Wrangham, W. (1901). Acute lead poisoning in women resulting from the use of diachylon as an abortifacient. *The British Medical Journal*, 13 June 1901.

WRc (1999). Possibility of using lining or coaring techniques for reducing exposure to lead from water supply pipes. DETR/DWI 4712.

Wycock, B.M., Schock, M.R. and Eastman, J.A. (1991). A study of the effects of municipal ion-exchange softening on the corrosion of lead, copper and iron in water systems. *Proceedings of the AWWA Annual Conference*.

Appendix 1 Case studies

A1.1 Brussels (BE): partial lead pipe replacement

Country	Belgium
Area of case history (town, city, etc.)	Brussels
Population in area	1.050.000
Year(s) of case history	2002 to 2008
Undertaken by	Vivaqua, Brussels
Publication reference (if available)	

Description of area:

Urban and suburban.

Water quality characteristics:

15% good quality surface derived water that receives comprehensive physico-chemical treatment.
85% good quality groundwater (only treated with chlorine) Corrosion inhibitors are not dosed.

Occurrence of lead pipes:

In 2002, Brussels Water Company estimated that they had 60,000 lead connection pipes in service.

Other known sources of lead in drinking water:

The number of lead pipes within houses is not known. Most internal pipes are galvanised iron.

Intervention measures undertaken:

The Water Company decided to replace all of its lead connection pipes over the period 2002 to 2013. By 2008, 38000 had been replaced. Replacement work should finish in 2011 as fewer lead connection pipes were found, compared to the original estimate.

At the time of replacing a lead connection pipe, the home owner was invited to replace their lead pipe at the same time.

Results of intervention measures:

From 2004 to 2008, 1280 random daytime samples were taken from houses where the lead connection pipe had been replaced:

1115 samples recorded <10 µg/l lead (87.11%)
94 samples recorded lead between 10 and 25 µg/l (7.34%)
71 samples recorded lead >25 µg/l (5.55%)

Random daytime sampling from across Brussels gave the following results:

	2004	2005	2006	2007	2008	TOTAL
Total samples	953	677	762	718	707	3817
Pb$<$ 10 µg/l	895	643	726	677	670	3611 (94.60%)
Pb 10–25 µg/l	43	27	22	30	28	150 (3.93%)
Pb$>$ 25 µg/l	15	7	14	11	9	56 (1.47%)

Conclusions:

The results indicate that the replacement of lead connection pipes has only been partially successful and that the replacement of lead pipes owned by home-owners is also required.

A1.2 Cambridge (UK): optimised ortho-phosphate dosing

Country	United Kingdom
Area of case history (town, city, etc.)	Cambridge
Population in area	250,000
Year(s) of case history	1995–2004
Undertaken by	Cambridge Water & WQM
Publication reference (if available)	Hayes *et al.*, 2006

Description of area: Cambridge City is located in Eastern England. The City is well known for its University and has surrounding suburban districts dating back to before the 1970s. The City and its suburban districts are surrounded by arable farm land and numerous villages.
Water quality characteristics: All water supplies to the Cambridge area are from groundwater abstractions, the majority being from a chalk aquifer, with a small amount abstracted from greensand. These waters have an alkalinity of 200 to 300 mg/l (as $CaCO_3$) and a pH of 7.0 to 7.5. Sulphate and chloride contents are low but nitrate approaches 50 mg/l (NO_3) at some abstractions.
Occurrence of lead pipes: Surveys indicated that 52% of houses in the City have a lead connection pipe (partly owned by the water company and the consumer) and that only 5% of houses have internal lead pipes.
Other known sources of lead in drinking water: None
Assessment of the extent of lead problems before intervention: Prior to the commencement of ortho-phosphate dosing in 1995, 68% of random daytime (RDT) samples complied with the lead standard of 10 µg/l.
Intervention measures undertaken: Ortho-phosphate dosing to the City's supplies commenced in 1995 at an average dose of 0.9 mg/l (P), geared to achieve the lead standard of 50 µg/l that applied at this time. Ortho-phosphate dosing was further optimised over the period 2000 to 2003, resulting in the application of an increased average dose of 1.2 mg/l (P). This average dose has been applied using a winter dose of 1.0 mg/l (P) and a summer dose of 1.5 mg/l (P).
Results of intervention measures: The earlier dosing of ortho-phosphate achieved 95% compliance with a lead standard of 10 µg/l, based on RDT sampling. Optimisation of ortho-phosphate dosing, at slightly higher doses, achieved >99% compliance with a lead standard of 10 µg/l, based on RDT sampling.
Conclusions: Optimised ortho-phosphate dosing achieved >99% compliance with a lead standard of 10 µg/l, based on RDT sampling.

A1.3 Den Haag (The Hague) (NL): pH elevation and lead pipe replacement

Country	Netherlands
Area of case history (town, city, etc.)	Den Haag (The Hague)
Population in area	450,000
Year(s) of case history	1985 to 2006
Undertaken by	Dunea (formerly Dune Water Works of South Holland
Publication reference (if available)	Van Dongen *et al.* (2008)

Description of area:

The drinking water in The Hague is supplied by Dunea from two of its treatment facilities: one at Scheveningen in the Northwest of the city and one in Monster to the South of the city. Dunea employs an artificial recharge system: pre-treated water from a dammed branch of the river Meuse, some 75 kilometres away from The Hague, is infiltrated into the coastal sand dunes. The water is abstracted after a residence time of, on average, about 60 days. After multi-stage treatment, the drinking water is fed into the distribution system of the city.

Water quality characteristics:

The water is conditioned by pellet softening in up-flow fluidized bed reactors: Scheveningen uses $Ca(OH)_2$ solution, whereas Monster uses NaOH solution. The hardness of the water recovered from the dunes has been reduced from 2.3 to 1.6 mmol/l (from 13 to 8.5 German hardness degrees and the pH has been raised from about 7.8 to about 8.5.

Occurrence of lead pipes:

Almost 100,000 lead connection pipes were present in The Hague, although the number of lead domestic installations is not known.

Assessment of the extent of lead problems before intervention:

In 1977, before lead replacement had begun and when there was no central conditioning, high lead contents were measured in flow proportional tap samples (median value was 65 μg/L).

Intervention measures undertaken:

Installation of a central conditioning plant at the Scheveningen water treatment facility (December, 1993), and later also at the treatment facility in Monster (2006).

The water supply company started with the removal of lead communication pipes in 1985, but only in cases where cast iron junctions were replaced. From 1994 onwards two consecutive projects were carried out to systematically remove lead communication pipes. This was the company's response to the tightening lead standard values in drinking water. The renovation of domestic drinking water installations was and remains the responsibility of the house-owner.

Results of intervention measures:
Central conditioning at Scheveningen resulted in the median values of lead in tap samples being reduced from 42–65 μg/L to about 25–27 μg/L Subsequent removal of lead connection pipes resulted in a mean value for the lead concentration in 2006 of 9.5 μg/L. However, 23% of the 2006 samples still had a lead content of > 10 μg/L. As the lead removal programme of Dunea had been completed by then, the high lead concentrations still measured must have been caused by remaining domestic installations with lead pipes. As a general observation, home owners were reluctant to replace their lead pipes because of the disruption and costs involved (even with Government grant-aid available).
Conclusions:
Central conditioning alone is not sufficient to reach a lead standard of 10 μg/L at the consumer's tap. Removal of lead connection pipes further lowered lead concentrations at consumers' taps, but not sufficiently to achieve 10 μg/l.

A1.4 Lisbon (PT): ortho-phosphate dosing strategy

Country	Portugal
Area of case history (town, city, etc)	Lisbon
Population in area	650,000
Year(s) of case history	2001 to 2003
Undertaken by	EPAL-Empresa Portuguesa das Águas Livres, S.A. and WQM – Water Quality Management Ltd
Publication reference (if available)	

Description of area:
There are four main sources of water supply: (i) Castelo de Bode (reservoir); (ii) Tagus (river); (iii) groundwater; and (iv) springs. Two treatment plants are installed which involves chemical coagulation, clarification, filtration, pH adjustment (8.0 to 9.0) and disinfection. These waters, after appropriate treatment, mix substantially on route to the City. Of a total production of 660 Mld, about 24% is supplied to the City, with the remainder making up bulk supplies to 35 surrounding municipalities North Tagus River. The distribution system in the City is divided into four zones, based on altitude.
Water quality characteristics:
The pH of supplies in the City is in the range 7.5 to 8.5, with alkalinity between 30 and 100 mg/l ($CaCO_3$). Organic content is low (TOC – 0.77–1.66 mg/L).

Occurrence of lead pipes:
Surveys undertaken by EPAL indicated that 26% of water supply connections in the City were made with lead piping, most typically either 25 or 38 mm internal diameter. The number of lead pipes within houses is not known.
Assessment of the extent of lead problems before intervention:
From 1999 to 2002, 17.5% of random daytime samples had lead >10 μg/l and 7.9% >25 μg/l. Laboratory plumbosolvency testing gave an average lead concentration after 30 minutes contact of around 50 μg/l Zonal modelling indicated that an average ortho-phosphate dose in the range 0.6 to 0.9 mg/l (P) should achieve >98% compliance with the future lead standard of 10 μg/l.
Intervention measures undertaken:
A pilot ortho-phosphate dosing trial was undertaken in one zone in 2003. A subsequent project then identified three dosing locations within the distribution network that could treat all of the City's water supplies. Approval from the Portuguese Government to proceed with the installation of dosing plants is awaited. EPAL is replacing all of its lead connection pipes and by now only 4.2% are installed. The renovation of domestic drinking water installations was and remains the responsibility of the house-owner.
Results of intervention measures:
A pilot phosphate dosing trial in 2003 resulted in an average reduction in lead of 76.6 to 6.9 μg/l, based on lead pipe rigs and 30MS sampling.
Conclusions:
The strategy for dosing ortho-phosphate to Lisbon's water supplies will provide comprehensive public health protection, once approval is given to proceed. Replacement of lead connection pipes has only been partially successful, with 6% of the samples having a lead content of >10 μg/L

A1.5 Ottawa, Ontario (CA): effect of pH

Country	Canada
Area of case history (town, city, etc.)	Ottawa, Ontario
Population in area	780,000
Year(s) of case history	2002–2006
Undertaken by	City of Ottawa Drinking Water Services
Publication reference (if available)	Douglas *et al.*, 2004; Douglas *et al.*, 2007a,b.

Description of area: An area designated as Old Ottawa South with a four-block portion of 394 homes. This area is served by 6 inch and 8 inch diameter unlined cast-iron water main pipes installed in 1910, having poor water circulation because of the occurrence of dead-ends in the affected portion and many homes are supplied by lead service connections.
Water quality characteristics: Surface water source (Ottawa River) with low alkalinity (25 mg/L) and total hardness (35 mg/L) but high organic content (DOC = 6.5 mg/L) and colour (30–35 TCU). Secondary disinfection is achieved with the use of monochloramine. Historically, the City of Ottawa has used CaO quicklime to raise the pH of treated water to 8.5 pH units but this practice resulted in increased turbidity and poor pH control. A new strategy was selected whereby sodium hydroxide and carbon dioxide to achieve a treated water pH of 9.2 and alkalinity >35 mg/L (as $CaCO_3$) were to be phased in from 2002–2003. The initial phase of changing to sodium hydroxide coincided with an increase in lead levels.
Occurrence of lead pipes: The City of Ottawa estimates there are 24,000 utility and up to 40,000 private lead service connections in the City. Many homes are supplied by lead service connections in Old Ottawa South but no estimate was provided.
Other known sources of lead in drinking water: It was generally accepted that lead service connections were the major source of lead for drinking water.
Assessment of the extent of lead problems before intervention: In August, 2002, a pregnant customer, suspecting her service connection was made of lead, had her water tested and found lead concentrations exceeded the Ontario Drinking Water Standard of 10 μg/L for flowing samples. Generally, there were few exceedances of the regulated lead levels in the City of Ottawa prior to this point.
Intervention measures undertaken: A drinking water advisory was issued for the affected area. The City of Ottawa investigation included large-scale sampling at 290 homes which found lead results were much higher than observed in recent years. The probable cause was determined to be nitrification in the affected area, which had reduced the target pH of 8.5 to pH 7.8–8.2 range. This decrease in pH, combined with warm water temperatures (22°C), effectively increased lead solubility. An increase to pH 9.2 was implemented and quickly resulted in lead concentrations below the regulated level of 10 μg/L (flowing).

Results of intervention measures:
After increasing the pH target in September 2002, lead levels sharply decreased below 10 μg/L (average of 6–8 μg/L). Subsequent monitoring continued to demonstrate system-wide compliance with the existing drinking water standard at that time. For the years 2002–2007, the average lead concentration for a flowing sample was 2.6 μg/L. Sampling undertaken in 2006 using the regulated and a modified protocol with stagnation showed lead concentrations ranging from 0.5–5.8 μg/L (mean of 1.8 μg/L) for flowing and <0.5–9.5 μg/L (for 30 min. standing samples, respectively. During that period, the summer samples tended to be higher than winter samples (but still compliant), due to nitrification.
Conclusions:
Nitrification leading to pH depression increased lead levels. Increasing the pH of the water supply resulted in consistently lower lead levels, in compliance with the regulation in place at the time, and control of nitrification.

A1.6 Vienna (AT): partial lead pipe replacement

Country	Austria
Area of case history (town, city, etc.)	Vienna
Population in area	1.690.936
Year(s) of case history	2003–ongoing
Undertaken by	Austrian Institute of Technology GmbH–AIT
Publication reference (if available)	–

Description of area:
Since 2003 a consumer-friendly "watercheck" has been offered to broad public in Austria. Lead in drinking water is not a major concern in Austria with the exception of some major city especially Vienna. Therefore one of the focal points of this study has been the determination of the extent of heavy metal leaching with special emphasis on lead within the 23 districts of Vienna because of multitude of buildings older than 60 years.
Water quality characteristics:
The Viennese drinking water comes from two main sources–primarily karstic spring waters and secondarily from a tertiary aquifer. Both sources show a low to medium mineralisation (electric conductivity: 160–390 μS/cm). The pH varies from 6.3 to 8 depending on the supply situation. Calcium (35–70 mg/l), magnesium (5–15 mg/l) and hydrogencarbonate (128–231 mg/l) form the major ions. The sulphate content ranges from 1.8–28 mg/l and the nitrate content from 2.5–9 mg/l. The sodium, the potassium and chloride content are very low (Na: <2.6 mg/l; K: <0.4 mg/l; Cl: <5.5 mg/l). All samples were taken from the tap within the costumer's house or apartment by the costumer.

Occurrence of lead pipes: Since the beginning of this study in 2003 several hundred meters of connecting pipes made of lead still existed. These pipes were replaced until 2007.
Other known sources of lead in drinking water: Rising mains made of lead within the houses and lead pipes within the apartments; brass fittings; faucets.
Assessment of the extent of lead problems before intervention: No published assessment available.
Intervention measures undertaken: Replacement of the remaining lead connecting pipes ($\ll$0.5% of the whole distribution network) until 2007.
Results of intervention measures: After 2007 the percentage of samples which are not meeting the required limit of 0.025 mg/l (Pb) lies still at 15 % although all connecting lead pipes were replaced.
Conclusions: The percentage of the samples, which show values over the limit, remained constant after 2007. This evidence indicates that the house installation (older buildings, even renovated, still have lead rising mains sometimes) play a decisive role in increasing the lead levels in drinking water at the tap.

A1.7 Wales (UK): optimised ortho-phosphate dosing

Country	United Kingdom
Area of case history (town, city, etc.)	Most of Wales
Population in area	3,000,000
Year(s) of case history	2002–2005
Undertaken by	Dwr Cymru Welsh Water & WQM
Publication reference (if available)	Hayes *et al.*, 2008

Description of area:

Wales is a devolved region in the west of the UK. The south was formerly an area of coal mining, ferrous and non-ferrous metal smelting and steel making; most of this industry has closed. The majority of Wales is rural with livestock farming in a hilly or mountainous terrain. The case study relates to the 29 major water supply systems that serve the region.

Water quality characteristics:

Most of the water abstracted for drinking water is surface derived from reservoirs and rivers, the majority being low in alkalinity and prone to colouration by natural organic matter. Treatment in most cases involves chemical coagulation, clarification, filtration, pH adjustment (8.0 to 9.0) and disinfection. In consequence, colour is removed and iron corrosion in distribution is minimal.

Occurrence of lead pipes:

The percentage of houses with a lead pipe varies from 8% to 65% with a regional average of 28%.

Other known sources of lead in drinking water:

It was generally considered that lead pipes were the major source of lead in drinking water in the region.

Assessment of the extent of lead problems before intervention:

The random daytime (RDT) sampling results before and after ortho-phosphate dosing in Wales are shown below:

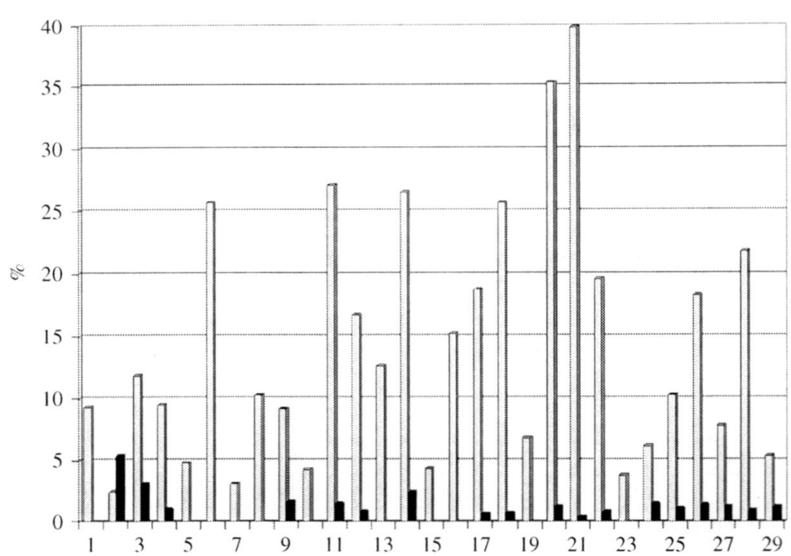

Key: light columns before dosing commenced, dark columns after

Intervention measures undertaken: In all cases, ortho-phosphate was dosed, mostly in the form of ortho-phosphoric acid. Doses ranged from 0.6 to 1.5 mg/l (P) with a regional average of 0.9 mg/l (P). Dose optimisation was achieved by an integrated approach involving increased RDT monitoring, the use of lead pipe test rigs at strategic locations, laboratory plumbosolvency testing, zonal modelling and tightening the operational control of ortho-phosphate dosing.
Results of intervention measures: On a regional basis, 99% compliance was quickly achieved with the future lead standard of 10 μg/l, based on RDT sampling.
Conclusions: Optimised ortho-phosphate has achieved a very high level of compliance with the future lead standard of 10 μg/l. The integrated approach to optimisation enabled the appropriate doses of ortho-phosphate to be achieved within a relatively short period.

Lightning Source UK Ltd.
Milton Keynes UK
27 March 2010

151983UK00001B/87/P

9 781843 393696